CAMBRIDGE EARTH SCIENCE SERIES

Editors: W. B. HARLAND (*General Editor*)
S. O. AGRELL, A. H. COOK, N. F. HUGHES

Jurassic Environments

'Ancient Dorset' by Henry de la Beche
This early and rather lurid portrayal of a Jurassic Environment seems to
anticipate by several decades the Darwinian 'Nature Red in Tooth and Claw'.

Jurassic Environments

A. HALLAM
University of Oxford

CAMBRIDGE UNIVERSITY PRESS

CAMBRIDGE

LONDON · NEW YORK · MELBOURNE

Published by the Syndics of the Cambridge University Press
The Pitt Building, Trumpington Street, Cambridge CB2 1RP
Bentley House, 200 Euston Road, London NW1 2DB
32 East 57th Street, New York, NY 10022, USA
296 Beaconsfield Parade, Middle Park, Melbourne 3206, Australia

© Cambridge University Press 1975

Library of Congress Catalogue Card Number: 74–80359

ISBN: 0 521 20555 7

First published 1975

Printed in Great Britain
at the University Printing House, Cambridge
(Euan Phillips, University Printer)

To the Memory of
WILLIAM JOCELYN ARKELL
(1904–1958)

Contents

Contents

ACKNOWLEDGEMENTS

Thanks are due to the following publishers and journals and to the authors
concerned for permission to reproduce the illustrations listed below:

The Geologists Association, figs. 2.1, 3.1; the Geological Society of
London, figs. 2.3, 7.8; the Society of Economic Palaeontologists and
Mineralogists, figs. 3.2, 3.3, 3.6, 5.5, 5.6, 5.7, 6.4; the Seel House Press,
figs. 2.2, 3.7, 8.1; the Palaeontological Association, figs. 3.8, 7.3; Elsevier
Publishing Company, figs. 4.8, 4.9, 4.10, 4.11; the Bayerische Akademie
der Wissenschaften, fig. 4.14; *Lethaia*, fig. 5.3; the American Association
of Petroleum Geologists, fig. 6.5; the Geological Society of America, figs.
6.6, 7.5, 7.6, 7.7; *Nature*, fig. 7.4; *New Zealand Journal of Geology and
Geophysics*, fig. 9.3.

Preface

In writing a relatively short book to conform with the requirements of the Cambridge Earth Science Monograph Series, one is inevitably faced with the problem of compression and of what to leave out. Much important work must be cited only briefly if at all and I would like to make at the outset the customary apology for being highly selective in my choice of material to mention. My selection should certainly not be held to reflect on papers that are merely cited in the reference list.

Among the numerous British and foreign colleagues who have helped me improve my knowledge of the Jurassic I must single out a few for special mention. I have benefited greatly from the consideration shown to me by several people who have taken the trouble to give me individual excursions to outcrops in other countries, namely Dr R. W. Imlay in the United States, M. R. du Dresnay in Morocco, Professor M. Misik in the Czechoslovakian Carpathians and Dr K. Seyed Emami in Iran. I also wish to thank Drs J. D. Hudson, H. C. Jenkyns and B. W. Sellwood and Mr W. B. Harland for critically reading various chapters, and Miss Helen Birch for typing the manuscript.

I feel my greatest debt, however, is to the late Dr W. J. Arkell, whose eloquent writings inspired me so much as a student and who must surely take his place along with Quenstedt, d'Orbigny, Oppel and Buckman as one of the leading students of Jurassic geology. I still have pleasant memories of the all too brief period before he suffered his first crippling stroke, when he was my research supervisor at Cambridge. Although his own work was primarily stratigraphic and taxonomic he was well aware that most of the future advances in Jurassic geology would come from the detailed analysis of facies. To people who knew him only distantly he could appear a somewhat stiff and forbidding figure, but I came to learn that this was merely the result of shyness. He was in fact very kind and generous to me, and the least I can do in return is to dedicate this book to his memory.

Oxford   A. HALLAM
October 1973

I Introduction

'The comprehensive Treatise of Geology has had its day.' With this trenchant sentence W. J. Arkell began his *Jurassic Geology of the World*. It is a measure of the proliferation of knowledge in the nineteen years since his celebrated work was published that we could say with equal force today that the comprehensive one-man treatise on a single geological system has also just about had its day.

The present, less voluminous work has more limited goals: to review and bring more into the public domain some of the advances in recent years in the interpretation of Jurassic facies and to apply the new knowledge and insights so obtained from this and other fields to the sorts of major topics which formed the subject matter of the concluding section of Arkell's book. The Jurassic researcher's necessary day-to-day preoccupation with details of stratigraphic or facies analysis should never be allowed to obscure the fact that our ultimate aim is to achieve as comprehensive and illuminating knowledge as possible of, for instance, the fluctuations in distribution of land and sea, the relationship of vulcanicity to plate tectonics, and the nature of the climate and the environmental control of biogeographic provinces.

Those who undertake to work on Jurassic facies often have the inestimable advantage of excellent stratigraphic control, because the zonation achieved by utilising ammonites compares favourably with that of almost any other fossil group in the Phanerozoic. Though this is not a book on stratigraphy as such, it was considered highly desirable to include a short chapter on the stratigraphic framework of environmental interpretation because it cannot justifiably be ignored in any comprehensive analysis of facies.

There follow four chapters dealing with litho- and biofacies analysis in varying degrees of detail for selected regions of Europe and North America. Knowledge of the facies, or even the basic stratigraphy, of other parts of the world is on the whole too sketchy and inadequate to warrant comparable treatment. Any book on the Jurassic is bound to show a strong bias towards Europe because this is where the great bulk of research has been done. There is no generally accepted way of subdividing the facies and

Introduction

the scheme adopted in chapters 3, 4 and 5 has an element of arbitrariness, as in most such cases, but has proved a convenient way of breaking down the subject matter.

The latter part of the book deals with more general topics, like the last section of *Jurassic Geology of the World*. This is, of course, only in the nature of an interim report on our present state of knowledge and the extent to which it is more than provisional can only be decided by future research. Of necessity much of the material has to be descriptive and non-quantitative and the interpretations relatively speculative. We still lack, for example, precise data on the temperatures and salinities of the seas, the nature of tides and ocean currents and the temperature, precipitation and wind patterns on land. It is also desirable to obtain better quantitative data on the distribution of particular types of sedimentary rocks and organisms and on changes of sea level. Perhaps we may look forward in the future to computer-assisted modelling of facies changes in space and time, and of climatic and ocean current patterns in a world whose disposition of continents and oceans was very different from the present day.

One of the most useful attributes of Arkell's book was its comprehensive list of references, and it would have been a great opportunity lost if a later work on the Jurassic had neglected to bring the reference list up to date. The volume of recent literature is enormous, and the following list of regional references cannot claim to be comprehensive. Most stratigraphic papers are only of limited importance and are appropriately published in local journals. In particular only a small proportion of the voluminous European literature is cited, but this should provide a helpful lead-in to the local literature. Where a series of papers exists on particular subjects only the more recent are cited, and pre-1955 papers are largely excluded, being adequately covered by Arkell's work. Palaeontological literature is ignored unless it has stratigraphic significance.

EUROPE

Austria. Blind (1963), Fabricius (1966), Fischer (1966), Fenninger (1967), Fenninger and Holzer (1970), Garrison and Fischer (1969), Jacobshagen (1965), Tollmann (1963).

British Isles. The English Midlands are dealt with in several chapters of the book edited by Sylvester Bradley and Ford (1968). A series of cyclostyled field guides to the classic regions of Dorset, north Somerset and Gloucestershire, east Yorkshire and the Scottish Hebrides were prepared for the William Smith Bicentenary meeting held in 1969 and can be obtained from a number of geological libraries in Britain. Scottish literature up to 1965 is reviewed in chapter 12 of *The Geology of Scotland* edited by G. Y. Craig.

Bate (1967), Berridge and Ivimey-Cook (1967), Callomon and Cope (1971), Donovan and Hemingway (1963), Edmonds *et al.* (1965), Falcon and Kent (1960), Green and Welch (1965), Howarth (1957, 1962, 1973), Howitt (1964), Kent (1967), Melville (1956), Morton (1965), Poole and Whiteman (1966), Sykes (1975), Taylor (1963), Terris and Bullerwell (1965), Wilson *et al.* (1958), Woodland (1971), Worssam and Ivimey-Cook (1971), Wright (1968, 1972).

Bulgaria. Chatalov (1967), Nachev (1966), Nachev *et al.* (1963), Sapunov (1959, 1969), Sapunov and Stephanov (1964), Sapunov *et al.* (1967).

Czechoslovakia. Androusev (1965), Misik (1964), Misik and Rakus (1964), Rakus (1964).

Denmark. Christensen (1963), Gry (1960), Larsen (1966), Nørvang (1957), Sorgenfrei and Buch (1964).

France. The 1961 BRGM publication *Colloque sur le Lias en France* contains a series of papers on the Lower Jurassic. There are numerous publications on French Jurassic stratigraphy, mostly only of local significance; only a limited selection of some of the more important is given here.
    Other useful references, including unpublished theses, are given by Mouterde *et al.* (1971 *a*).

Ager and Evamy (1963), Ager and Wallace (1966), Blaison (1961), Bonte *et al.* (1958), Bouroullec and Deloffre (1969), Cariou (1966), Carozzi *et al.* (1972), Donze (1958), Donze and Enay (1961), Elmi (1967), Elmi and Mouterde (1965), Enay (1966), Fischer (1969), Gabilly (1964*a*, *b*), Guex (1972), Hölder and Ziegler (1959), Pavia (1971), Schirardin (1955), Sturani (1966).

Germany (Bundesrepublik Deutschland). A comprehensive reference list up to 1964 is given in Hölder's book *Jura* and the classic Swabian Jurassic is amply dealt with in the guidebook by Geyer and Gwinner (1968).

Barthel (1969), Bottke *et al.* (1969), Brand and Hoffmann (1963), Buck *et al.* (1966), Herrmann (1971), Hoffmann (1966), Hoyer (1965), Urlichs (1966), Weber (1964), Zeiss (1968).

Introduction

Germany (*Deutsches Demokratisches Republic*). Ernst (1967), Rusbult and Petzke (1964), Schumacher and Sonntag (1964).

Greece. Aubouin (1959), Aubouin *et al.* (1962), Bernoulli and Renz (1970), Dercourt (1964), Renz (1955).

Hungary. Fülöp (1971), Galacz and Vörös (1972), Geczy (1961, 1966).

Norway and Spitsbergen. Ørvig (1960), Parker (1967).

Italy. Barbera Lamagna (1970), Bosellini and Broglio-Loriga (1971), Cantaluppi and Montanari (1969), Christ (1960), Colacicchi and Praturlon (1965), Crescenti (1971), D'Argenio (1967), D'Argenio and Scandone (1971), Fantini Sestini (1962), Farinacci (1967), Merla (1951), Montanari and Crespi (1974), Passerini (1964), Pinna (1966), Sirna (1962), Sturani (1962, 1964 *a, b*), Wendt (1964), Zanzucchi (1963).

Poland. The excellent Atlas of facies maps (Dadlez *et al.* 1964) and the various papers published in the 1st Jurassic Colloquium in Poland (*Inst. Geol. Warsaw, Biul.* **203** (1967)) serve as a valuable introduction. The Warsaw Geological Institute has also published a catalogue of Polish Jurassic fossils (1970).

Bielecka (1960), Birkenmajer and Znosko (1955), Dadlez (1964), Dayczak-Calikowska (1967), Jurkiewiczowa (1967), Karaszewski (1962), Kopik (1962), Kopik *et al.* (1967), Kutek (1961, 1962, 1967), Kutek and Glazek (1972), Ksiazkiewicz (1956), Malinowska (1967, 1972), Teofilak-Maliszewska (1967), Wierzbowski (1966).

Portugal. Hallam (1971 *c*), Mouterde (1967), Mouterde *et al.* (1971 *b*), Romariz (1960), Ruget Perrot (1962).

Rumania. Barbulescu (1971), Mutihac (1971), Patrulius and Popa (1971), Raileanu *et al.* (1964), Semaka (1961).

Spain. A symposium volume with numerous papers on the Jurassic stratigraphy of Spain has recently been published by the Instituto de Geologia Economica, Univ. Madrid (*Cuad. Geol. Iberica* 2, 1971).

Barthel *et al.* (1966), Behmel (1970), Behmel and Geyer (1966), Beuther *et al.* (1966), Busnardo *et al.* (1965), Dubar and Mouterde (1957), Dubar *et*

4

al. (1960, 1971), Geel (1966), Geyer (1967a), Paquet (1969), Querol (1969).

Sweden. Norling (1972), Reyment (1959), Vossmerbäumer (1970).

Switzerland. Bernoulli (1964), Donovan (1958), Grasmück (1961), Gross (1965), Gygi (1969), Jung (1963), Trümpy (1960, 1971), Wiedenmeyer (1963), Ziegler (1956).

USSR. The enormous Russian literature poses a serious problem for those who do not read the language because it is not customary even to publish English summaries. Fortunately there are several general, though brief, reviews in English or French, namely Beznozov *et al.* (1968), Gorsky and Leonenok (1964), Krimholz (1964) and Saks and Strelkov (1961). The bilingual Atlas of Russian Jurassic facies maps (Vereshchagin and Ronov, 1968) gives a wealth of data on the stage-by-stage distribution of sediments and fossils, with corresponding palaeogeographic interpretation. The most up-to-date comprehensive Russian language reference is the book edited by Krimholz (1972). Another recent book, edited by Saks (1972), deals with the Jurassic–Cretaceous boundary in the Boreal Realm.

Yugoslavia. Andelković (1966), Aubouin *et al.* (1965), Gusić *et al.* 1971, Pektović *et al.* (1960), Radoičić (1966), Veselinović (1963), Ziegler (1963).

AFRICA

Algeria. Bertraneau (1955), Cruys (1955), Elmi (1971), Emberger (1960).

Ethiopia. Clift (1956), Jordan (1971b).

Gabon. Hourcq and Reyre (1956), Krömmelbein and Wenger (1966).

Kenya. Cox (1965), Pulfrey (1963), Saggerson and Miller (1957), Thompson and Dodson (1960).

Libya. Desio *et al.* (1960, 1963).

Madagascar. Besairie and Collignon (1956), Blaison (1963), Collignon (1957, 1958–60, 1964a, b), Collignon *et al.* (1959).

Morocco. Agard and du Dresnay (1965), Ambroggi (1963), Blumenthal *et al.* (1958), Choubert and Faure Muret (1962), Colo (1961), Dubar (1962),

Introduction

Du Dresnay (1963, 1964 a, b), Durand Delga and Fallot (1957), Griffon and Mouterde (1964), Guex (1973), Martinis and Visintin (1966), Rouselle (1965), Soc. Chérif. Petrol. (1966).

Senegal. Spengler *et al.* (1966).

South Africa. Dingle and Klinger (1972), Klinger *et al.* (1972).

Tanzania. Aitken (1961), Cox (1965), Kent *et al.* (1971).

Tunisia. Bonnefous (1967), Bonnefous and Rakus (1965), Busson (1967).

ASIA

Afghanistan. Benda (1964), De Lapparent *et al.* (1966), Desio *et al.* (1965), Mensink (1967), Rossi Ronchetti and Fantini Sestini (1961).

Indonesia and Malaysia. Liechti *et al.* (1960), Wolfenden (1965).

India. Agrawal (1956), Mahadevan and Srivamadas (1958), Srivastava (1963).

Iran. Assereto (1966), Assereto *et al.* (1968), Fantini Sestini (1966), Huckriede *et al.* (1962), James and Wynd (1965), Seyed Emami (1971), Stöcklin (1968), Stöcklin *et al.* (1965).

Iraq. Dunnington *et al.* (1959).

Israel. Derin and Reiss (1966), Derin and Gerry (1972), Farag (1959), Friedman *et al.* (1971), Goldberg and Friedman (in press), Hudson (1958), Maync (1966), Raab (1962).

Japan. Hayami (1961), Hirano (1973), Minato *et al.* (1965), Sato (1962, 1964), Sato *et al.* (1963).

Jordan. Wetzel and Morton (1958).

Oman. Hudson and Chatton (1959), Morton (1959).

Pakistan. Davies and Gardezi (1965), Fatmi (1972), Williams (1959).

Phillippines. Andal *et al.* (1968), Gervasio (1967), Kobayashi (1957).

Saudi Arabia and Yemen. Greenwood and Bleackley (1967), Imlay (1970), Powers (1962), Powers *et al.* (1966).

Syria and Lebanon. Daniel (1963), Dubertret (1963), Haas (1955).

Thailand. Kobayashi (1960), Komalarjun and Sato (1964).

Turkey. Bremer (1965), Enay *et al.* (1971), Hölder (1964).

USSR. See Europe.

AUSTRALASIA AND ANTARCTICA

Antarctica. Ballance and Watters (1971), Borns and Hall (1969), Gunn and Warren (1962), Harrington (1965), Howarth (1958), Norris (1965), Plumstead (1964), Quilty (1970).

Australia. De Jersey (1960), Dettmann (1963), Guppy *et al.* (1958), McWhae *et al.* (1958), Playford (1959), Playford and Dettman (1965), Skwarko (1970), Veevers and Wells (1961).

New Guinea. Australian Petroleum Company (1961), Gerth (1965), Visser and Hermes (1962), Westermann and Getty (1970).

New Zealand. Fleming (1970), Stevens (1968).

NORTH AMERICA

Canada. A comprehensive review of the Jurassic of British Columbia, Alberta and the southern Yukon is given by Frebold and Tipper (1970). The Canadian Arctic ammonite faunas are dealt with by Frebold (1961, 1964). Other references include Gussow (1960), Jeletsky (1965, 1966), McIver (1972) and Roddick *et al.* (1966).

Greenland. Callomon (1959, 1970), Callomon *et al.* (1972), Donovan (1957, 1964), Surlyk and Birkelund (1972), Surlyk *et al.* (1973).

Mexico and Cuba. Alencaster de Cserna (1963), Cantu Chapa (1963), Erben (1956a, 1956b, 1957b), Hallam (1965), Khudoley and Meyerhoff (1971), Krömmelbein (1956, 1962), Meña Rojas (1960), Peña Muñoz (1964), Perez Ibarguengoitia (1965), Viniegra (1971), Verma and Westermann (1973).

Introduction

USA. A good general account of the Western Interior is given by Imlay (1957) and a series of facies maps for this region has been published by the US Geological Survey (McKee *et al.* 1956). Imlay has written a whole series of papers on US Jurassic faunas, only a selection of which are indicated here. Brenner and Davies (1974), Dickinson (1962), Hallam (1965), Harshbarger *et al.* (1957), Imlay (1956, 1957, 1962, 1964, 1967 *a*, *b*), Imlay and Detterman (1973), Imlay *et al.* (1959), Moberley (1960), Peterson (1957, 1972), Pipiringos (1968), Stanley (1971), Stanley *et al.* (1971), Tanner (1965), Thomas and Mann (1966), Westermann (1964, 1969).

SOUTH AMERICA

Argentina. Hillebrandt (1973 *a*, *b*), Stipanicic (1966, 1969), Stipanicic and Reig (1956), Stipanicic and Rodrigo (1970), Westermann (1967).

Chile. Biese (1957 *a*, *b*), Cartagena (1965), Carter (1963), Cecioni (1961), Cecioni and Garcia (1960), Cecioni and Westermann (1968), Corvalán Diaz (1957, 1959), Harrington (1961), Hillebrandt (1970, 1973 *a*,*b*), Klohn Giehm (1960).

Colombia. Bürgl (1965, 1967), Geyer (1967 *b*, 1968).

Peru. Rüegg (1957), Schindewolf (1957).

2 The stratigraphic framework

The Jurassic System is of exceptional importance in the study of stratigraphy because many of the basic principles and concepts were first enunciated after study of its rocks and fossils in Europe. Thus William Smith formulated his 'Law' of Superposition and first recognised the utility of fossils in correlation, d'Orbigny erected the first set of stages, meant to be valid over the whole world, and Oppel was the founder of modern zonal stratigraphy. The classic work of these and other pioneers is amply reviewed by Arkell (1933) and Hölder (1964) and we need only draw attention here to modern developments.

The fundamental biostratigraphic unit is, of course, the zone, normally defined by the best stratigraphic guides, ammonites, but the term has been used in a variety of ways and is a potential source of confusion. In contrast to his lengthy discussion in 1933 of such varied terms as biozones, faunizones, teilzones, epiboles and hemerae, Arkell's stratigraphic introduction to his *Jurassic Geology of the World* (1956) dealt only briefly with formations, stages and zones, the other terms being considered more or less redundant.

It is clear that in its normal usage by Jurassic stratigraphers, the zone is a *local-range biozone* as defined by the Geological Society of London committee on stratigraphic nomenclature (Harland *et al.* 1972). In practice the range may be very extensive at times when particular ammonite faunas were widespread and did not exhibit marked provinciality, as for instance in the early part of the period. When the same sequence of ammonite faunas is found over extensive parts of the world, as appears to be quite often the case at generic and even species level, we may assume with confidence that the defined zones correspond essentially to chronological divisions, for the following reason. If given ammonite genera or species originated in different areas at different times and also exhibited individual patterns of migration, at different speeds and directions, we should expect to find no consistent sequence in different areas. If a sequence of lower taxa $A \ldots N$ were found to occur in, say, different continents we should be forced to conclude, if chronological contemporaneity were to be denied, that the migration rates and direction for a whole succession of ammonites were identical over millions of years, which seems highly implausible. As the latter situation is generally the case, with the same sequence occurring in different areas of the same faunal province, homotaxis on a significant scale can be ruled out.

Not much progress has yet been made in translating biostratigraphic units to the standard stratigraphic scale, with the establishment of 'bench marks' in chosen rock successions to define type stages and chronozones, in the manner recommended by Harland *et al.* (1972). Stratigraphic principles and procedures have been fully discussed at several international conferences on the Jurassic, the most notable of which were those organised by P. L. Maubeuge and held in Luxembourg in 1962 and 1967. (The first of these was published as '*Colloque du Jurassique*', *Publ. Inst. Grand-Ducal, Sect. Sci. Nat., Phys., Math., Luxembourg*, 1964, and the second as '*Colloque du Jurassique à Luxembourg*', Editions du Bureau de Recherche Géologique et Minière, Paris, 1973.) Some of the recommendations made at these various conferences bear a distinctly national stamp. For the time being any such recommendations must be treated as provisional. Among the most useful in practice have been those of Callomon (1965) and accordingly they will be dealt with here in some detail.

Callomon distinguished *typological definitions*, in which stratigraphic units are defined by type sections, from *hierarchical definitions*, in which units are accorded *rank* and a unit of higher rank is defined in terms of the lower rank which it encompasses. It is considered desirable to have a principle of priority in order to create stability of nomenclature, and a typological definition at one stage in the hierarchy to provide an objective anchor to the whole structure.

It is proposed that *standard zones* be erected, based on a type section in which specified beds yield a characteristic fauna. The base of the lowest bed defines the base of the zone and the top of the zone is defined by the base of the succeeding zone. The name given to the zone is that of a characteristic or common species but the zone's definition should be independent of whichever species is chosen as the index, or of the exact total range of any one species (or biozone). The stratigraphical extent of biozones is likely to vary from time to time, as more research is undertaken, and hence is unstable; there is also likely to be overlap with other biozones The name should be written as, for instance, *Macrocephalus Zone* (of Oppel) or zone of *Macrocephalites macrocephalus* (a Schlotheim species). Locally it may be possible to subdivide the succession further into *subzones* or even *horizons*, again labelled by a characteristic fossil.

Once the zones are defined typologically, the stages may be simply defined in terms of these zones. Definition in terms of stage stratotypes, as favoured by many French workers, can only lead to chaos. For example, the 'type sections' of the Toarcian (Thouars) and Bajocian (Bayeux) actually overlap, and the base of neither as originally defined coincides with the type section of the lowest zone of either stage.

Considerable progress has already been made by British workers in proposing suitable type sections for many classic zones, but regrettably it cannot yet be said that Calloman's recommendations have received wide international recognition.

The subdivision of the Jurassic into Lower, Middle and Upper units has been the subject of continuing controversy. Although the Russians still take the Callovian stage as the base of the Upper Jurassic, a consensus is developing which supports Arkell's (1956) proposal that it is better put in the Middle Jurassic. This indeed accords more closely with von Buch's classic subdivision of the German strata into Black Jurassic (*Schwarze Jura* or Lias), Brown Jurassic (*Braun Jura* or Dogger) and White Jurassic (*Weiss Jura* or Malm). Although Arkell recommended that Aalenian should be dropped as a stage term it has continued to be widely used in Europe. Since it can be unambiguously defined as corresponding exactly to Arkell's Lower Bajocian it seems worth adopting as one of the standard stages. Although it has in the past sometimes been put in the Lower Jurassic there is now general agreement following the Luxembourg conferences that it should mark the base of the Middle Jurassic.

German stratigraphers have continued to use Quenstedt's sixfold division of each of the three major units, the divisions being labelled with the Greek letters alpha to zeta. These units have a status more or less intermediate between a zone and a stage. Although they are well understood by stratigraphers working on their local German successions their widespread use obstructs understanding and should be discouraged. The reader wishing to translate the Quenstedt units to stages and zones should consult Hölder (1964).

STAGE AND ZONAL SUBDIVISIONS

The stage terms adopted in this book are given in table 2.1. Since Europe is classic terrain for Jurassic stratigraphy it is desirable also to give a list of currently recognised zones. Discussion on the zonation need only be brief, and will concentrate upon departure from the scheme put forward by Arkell (1956). A Geological Society of London stratigraphic subcommittee (George *et al.* 1969) has made proposals about how to define the base of the stages from the Hettangian to the Kimmeridgian in terms of subzones or chronozones at specific European localities.

TABLE 2.1   *Jurassic stages*

UPPER	TITHONIAN–VOLGIAN	
	KIMMERIDGIAN	
	OXFORDIAN	
MIDDLE	CALLOVIAN	
	BATHONIAN	
	BAJOCIAN	
	AALENIAN	
LOWER	TOARCIAN	
	PLIENSBACHIAN	
	SINEMURIAN	
	HETTANGIAN	

TABLE 2.2   *Lower Jurassic stages and zones*

Stages	Substages	Zones
TOARCIAN	UPPER TOARCIAN or YEOVILIAN	*Dumortieria levesquei* *Grammoceras thouarsense* *Haugia variabilis*
	LOWER TOARCIAN or WHITBIAN	*Hildoceras bifrons* *Harpoceras falciferum* *Dactylioceras* *tenuicostatum*
PLIENSBACHIAN	UPPER PLIENSBACHIAN or DOMERIAN	*Pleuroceras spinatum* *Amaltheus margaritatus*
	LOWER PLIENSBACHIAN or CARIXIAN	*Prodactylioceras davoei* *Tragophylloceras ibex* *Uptonia jamesoni*
SINEMURIAN	UPPER SINEMURIAN	*Echioceras raricostatum* *Oxynoticeras oxynotum* *Asteroceras obtusum*
	LOWER SINEMURIAN	*Caenisites turneri* *Arnioceras semicostatum* *Arietites bucklandi*
HETTANGIAN		*Schlotheimia angulata* *Alsatites liasicus* *Psiloceras planorbis*

Hettangian to Toarcian

The zonal scheme for the Liassic of northwest Europe proposed in a lengthy paper by Dean *et al.* (1961) has proved widely acceptable and warrants little comment (table 2.2). It differs from Arkell's in the addition of the Liasicus Zone to the Hettangian and the much-needed subdivision of the Upper Toarcian, which was formerly accorded only one zone. The scheme more recently proposed for France by Mouterde *et al.* (1971 *a*) differs in a few respects. Thus the Rotiforme Subzone (Bucklandi Zone) of Dean *et al.* is elevated to a zone, as is the Stokesi Subzone (Margaritatus Zone). A more refined zonal subdivision of the Upper Toarcian is also attempted. The upper part of the Thouarsense Zone, marked by the appearance of *Pseudogrammoceras*, is separated as the zone of *Hammatoceras insigne*, and the Levesquei Zone is subdivided into a lower, Pseudoradiosa and upper, Aalensis Zone.

Based upon his collecting in the Toarcian of Italy and southern Switzerland, where the northwest European zonation breaks down, Donovan (1958) proposed three zones and six subzones. The southern and northern European ammonite faunas coexist in Portugal, allowing a tentative correlation to be made (Hallam 1971 *c*). Donovan's Mercati Zone appears to correspond with the Bifrons Zone, and his Erbaense and Meneghinii zones respectively with the Variabilis–Thouarsense and Levesquei zones. No zonal subdivision for the Pliensbachian of southern Europe as a whole has yet been satisfactorily achieved, though Mouterde's (1967) work in Portugal is the most promising in this aspect. Such a subdivision is badly needed because the amaltheid ammonites on which the Dean *et al.* scheme is based are rare or absent in the circum-Mediterranean region and the affinities of the ammonite fauna are markedly Tethyan.

Aalenian to Callovian

The northwest European zonal scheme of table 2.3 differs only slightly from that of Arkell. His Lower Bajocian becomes the Aalenian and his Middle Bajocian the Lower Bajocian. Arkell's Scissum Zone, inserted between the Opalinum and Murchisonae zones, is of little utility outside England because the zonal index *Tmetoceras scissum* ranges from the Opalinum to the top of the Murchisonae Zone in France (Mouterde *et al.* 1971 *a*). On the other hand it seems useful, following the French workers, to split off the Concavum as a separate zone. Zonation of the Lower Bajocian has for a long time been known to be unsatisfactory and has now been revised by Parsons (1974). The zone of *Sonninia sowerbyi* cannot be used because the index species was incorrectly identified by Oppel and later workers and the type

TABLE 2.3 *Middle Jurassic stages and zones*

Stages	Substages	Zones
CALLOVIAN	UPPER	*Quenstedtoceras lamberti*
		Peltoceras athleta
	MIDDLE	*Erymnoceras coronatum*
		Kosmoceras jason
	LOWER	*Sigaloceras calloviense*
		Macrocephalites macrocephalus
BATHONIAN	UPPER	*Clydoniceras discus*
		Oxycerites aspidoides
		Prohecticoceras retrocostatum
	MIDDLE	*Morrisiceras morrisi*
		Tulites subcontractus
	LOWER	*Procerites progracilis*
		Zigzagiceras zigzag
BAJOCIAN	UPPER	*Parkinsonia parkinsoni*
		Garantiana garantiana
		Strenoceras subfurcatum
	LOWER	*Stephanoceras humphriesianum*
		Emileia (Otoites) sauzei
		Witchellia laeviuscula
		Hyperlioceras discites
AALENIAN		*Graphoceras concavum*
		Ludwigia murchisonae
		Leioceras opalinum

specimen is quite indeterminate. It is replaced by the zone of *Witchellia laeviuscula*, and the subzone of *Hyperlioceras discites* is elevated to a zone; the zone of *Emileia (Otoites) sauzei* is redefined.

The Bathonian zonation (fig. 2.1) is that of Torrens (1965, 1969) and has proved of widespread applicability in Europe; it represents a considerable improvement on that of Arkell. The scheme of Mouterde *et al.* differs in that only four zones are recognised, successively Zigzag, Subcontractus, Retrocostatum and Discus. Callomon (1959) has proposed a sequence of seven zones in East Greenland ranging in age from Upper Bajocian to the top of the Bathonian (based on the genera *Cranocephalites*, *Arctocephalites*, *Arcticoceras* and *Cadoceras*). Precise correlation with Europe is impossible because the Greenland ammonites belong to the Boreal Realm, but subsequent work indicates that the Greenland zones may be widely applicable in the Arctic.

In the French scheme the Calloviense Zone of the Callovian is replaced

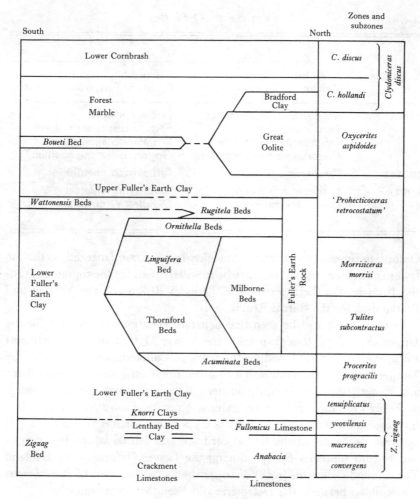

FIGURE 2.1 Zonal subdivision of the Bathonian formations of southern England (Torrens 1969).

by the zone of *Macrocephalites gracilis* because *Sigaloceras calloviense* is rare or absent in France.

Oxfordian to Tithonian and Volgian

Although Tethyan and Boreal provinciality in the ammonite faunas is recognisable in both Lower and Middle Jurassic faunas there are enough fossils common to northern and southern Europe to allow good correlation. The problem of correlation becomes much more serious from Upper

TABLE 2.4 *Oxfordian*

| Substages | Zones | |
	N.W. European Province (Britain, N.W. Paris Basin)	Submediterranean Province (S. and E. Paris Basin, S. France)
UPPER	*Ringsteadia pseudocordata*	*Idoceras planula*
	Decipia decipiens	*Epipeltoceras bimammatum*
	Perisphinctes cautisnigrae	*Perisphinctes bifurcatus*
	Gregoryceras transversarium	*Gregoryceras transversarium*
MIDDLE	*Perisphinctes plicatilis*	*Perisphinctes plicatilis*
LOWER	*Cardioceras cordatum*	*Cardioceras cordatum*
	Quenstedtoceras mariae	*Quenstedtoceras mariae*

Oxfordian times onwards as provinciality became more marked, so that no fewer than three stage terms have been widely used for the topmost part of the Jurassic, the Tethyan for the Tethyan Realm and the Volgian and Portlandian for the Boreal Realm.

In table 2.4 it will be seen that separate zonations are needed for the Upper Oxfordian, though not for the Lower Oxfordian. The northwest European zonation is based on the work of Callomon and is accepted by Mouterde *et al.* for northwestern France. Their zonation for the southern and eastern Paris Basin and southern France is quite different, and is probably valid for a much more extensive region of southern Europe. The increased difficulty of correlating the Upper Oxfordian is a consequence of northwards retreat of the boreal cardioceratid faunas which had ranged widely into southern Europe during the Lower Oxfordian. In England Torrens and Callomon (1968) distinguished a further zone of *Perisphinctes variocostatus* between the Decipiens and Pseudocordata zones.

The index fossil of the Transversarium Zone is generally rare but the associated fauna allows reasonably precise correlation. Correlation of the Bifurcatus and Cautisnigrae zones is based on the work of Enay (1966) in the French Jura. The Bimammatum Zone is characterised by the appearance of *Epipeltoceras* and the Planula Zone by the abundance of *Idoceras*. *Decipia* is common to both regions and hence facilitates correlation of the Decipiens Zone.

The stage term Kimmeridgian has been used in different senses by British and continental workers. The Kimmeridgian *sensu gallico* is equivalent only to the Lower Kimmeridgian *sensu anglico*. Since the former is followed by the Tithonian and the redefined Volgian there is a strong case for adopting it widely. Once more the marked faunal differentiation demands separate zonal schemes for northwestern and southern Europe (table 2.5).

TABLE 2.5   *Kimmeridgian* (sensu gallico)

N.W. Europe	S. Europe
Aulacostephanus autissiodorensis	*Hybonoticeras beckeri*
Aulacostephanus eudoxus	*Aulacostephanus eudoxus*
Aulacostephanus mutabilis	*Aspidoceras acanthicum*
	Crussoliceras divisum
Rasenia cymodoce	*Ataxioceras hypselocyclum*
Pictonia baylei	*Sutneria platynota*

TABLE 2.6   *Tithonian*

Substages		Zones

UPPER		*Paraulacosphinates transitorius*
		  (*Berriasella delphinensis* and *chaperi*)
		Pseudovirgatites scruposus
		  (*Micracanthoceras micracanthum*)

LOWER
sensu lato
	Middle	*Pseudolissoceras concorsi*
		  (*P. bavaricum*)
		Sublithacoceras penicillatum
		  (*Virgatocimoceras rothpletzi*)

	Lower *sensu stricto*	*Franconites vimineus* *Dorsoplanitoides triplicatus* *Glochiceras lithographicum* (*Hybonoticeras hybonotum*)

The upper part of the Kimmeridgian is characterised by *Aulacostephanus*. Since the index species is absent in a large part of Europe, Ziegler (1961) has proposed replacing the Pseudomutabilis by the Eudoxus and Autissiodorensis zones. *A. eudoxus*, being common to both regions, provides a useful tie-up. *A. mutabilis* is also found in the Acanthicum Zone, and *Rasenia* species in the Platynota Zone.

The base of the Tithonian is defined by *Gravesia*, which extends into both England and the Volga Basin and hence provides an excellent means of correlation. Above this horizon the situation becomes very confused, and much more work is needed to resolve the many problems involved in establishing both interprovincial correlation and stratigraphic precision.

The stratigraphic framework

Consequently the scheme proposed below must be regarded as tentative and provisional.

The French workers proposed a zonal scheme which they believed to be valid for the classic Lower Tithonian of Bavaria (Zeiss 1968) and Upper Tithonian of the Ardèche region of southern France (table 2.6). It will be seen that there is still uncertainty about whether to subdivide the stage into two or three major units. Correlation with Arkell's 'Middle Tithonian' zone of *Semiformiceras semiforme* is not yet well established. The Palmatus–Ciliata 'zones' of Arkell are thought to be equivalent to the Concorsi Zone. Subdivision of the Upper Tithonian has proved especially difficult and at present is probably done more precisely on calpionellids rather than ammonites.

For the type Volgian of the Russian Platform, Gerasimov and Mikhailov (1966) have recently revised the zonal classification. The zonal sequence is given in table 2.7.

TABLE 2.7   *Volgian*

Substages	Zones
UPPER	*Craspedites nodiger* *Craspedites subditus* *Kachpurites fulgens*
MIDDLE	*Epivirgatites nikitini* *Virgatites virgatus* *Dorsoplanites panderi*
LOWER	*Subplanites pseudoscythicus* *Subplanites sokolovi* *Subplanites klimovi*

The scheme of table 2.7 differs from the traditional one in that the base of the Volgian is brought down to a horizon containing *Gravesia*. The new Lower Volgian thus corresponds to the upper part of the English Kimmeridge Clay and the old Lower Volgian becomes the Middle Volgian. The zone of *Riasanites rjasanensis* is transferred from the top of the Jurasic into the basal Cretaceous (Ryazanian stage). By lowering the previously defined boundary of the base of the Volgian Gerasimov and Mikhailov offended against the principle of priority recommended by Callomon. On the other hand, their scheme has the considerable advantage of allowing a precise correlation to be made between the bases of the Tithonian and Volgian. For this reason it seems worth adopting.

With regard to the base of the Cretaceous in the Tethyan and Boreal

Realms, the association in the Caucasus of the species *Riasanites rjasa-nensis* and *Thurmanniceras boisseri* suggests that the Ryazanian and Berriasian will correlate only if the bottom third of the type Berriasian is transferred to the Tithonian (Casey 1963).

A new zonation is also proposed for the Upper Kimmeridgian *sensu anglico*. Based upon his collecting in Dorset, Cope (1967) dropped the *Gravesia* zones of the basal part because the genus is rare, and erected the following zones, all with index species of the genus *Pectinatites; Gravesia* occurs in the bottom two zones:

> Pectinatus
> Hudlestoni
> Wheatleyensis
> Scitulus
> Elegans

Casey (1967) reversed the order of the names of the two *Pavlovia* zones at the top of the Kimmeridge Clay in southern England, so that the Rotunda comes to succeed the Pallasioides Zone. He also suggested transferring these zones from the Kimmeridgian *sensu anglico* to the Portlandian. The Portlandian stage would thus embrace both the Middle Volgian of Gerasimov and Mikhailov and the Portland Beds of southern England. He criticised the correlation between the type Volgian and the Portland Beds, suggested by the two Russian workers, which is based upon the supposed common occurrence of *Pavlovia, Zaraiskites, Crendonites* and *Kerberites*. He questioned the identification of the Russian supposed *Crendonites* and *Kerberites* and argued that the Portland Beds were deposited at a time of non-deposition between the Middle and Upper Volgian, when the sequence is generally admitted to be highly condensed, with non-sequences.

Casey (1962) was also responsible for demonstrating the presence of Upper Volgian deposits in central England when he recognised craspeditids in the lower part of the Spilsby Sandstone, hence returning to the correlation proposed earlier this century by Pavlov. The Upper Volgian in southern England is presumably represented by the lower part of the non-marine Purbeck Beds.

Since the Jurassic of southern England passes into Cretaceous in largely non-marine facies the problem arises of where to place the boundary. Casey (1963) proposed that the distinctive marine horizon of the Cinder Bed in the Middle Purbeck Beds of Dorset be taken as this boundary. This is because the horizon can be traced northwards into strata with a more varied molluscan fauna, and appears to correlate with the upper part of the Serpulite Formation of northern Germany, whose equivalent in Poland

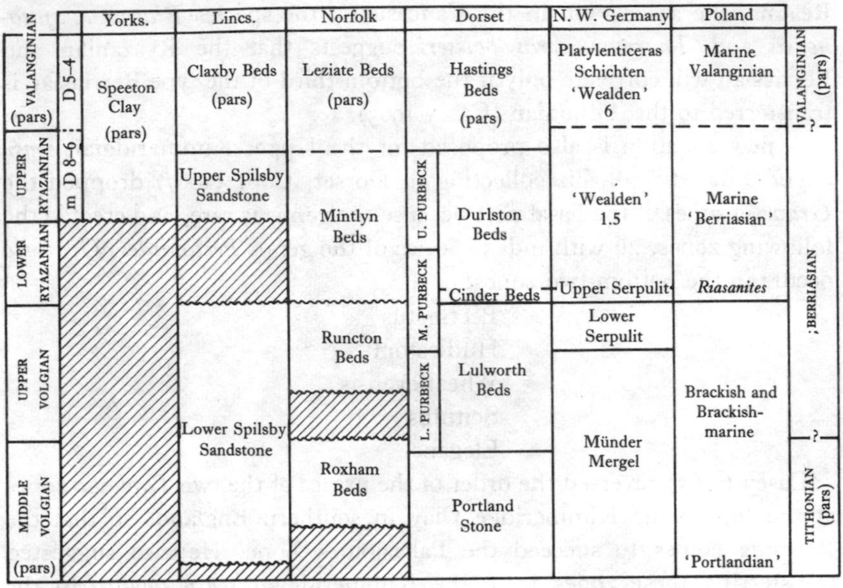

		Yorks.	Lincs.	Norfolk	Dorset	N. W. Germany	Poland		
VALANGINIAN (pars)	D 5-4	Speeton Clay (pars)	Claxby Beds (pars)	Leziate Beds (pars)	Hastings Beds (pars)	Platylenticeras Schichten 'Wealden' 6	Marine Valanginian	VALANGINIAN (pars)	?—
UPPER RYAZANIAN	m D 8-6		Upper Spilsby Sandstone	Mintlyn Beds	U. PURBECK — Durlston Beds	'Wealden' 1.5	Marine 'Berriasian'	BERRIASIAN	
LOWER RYAZANIAN					Cinder Beds —	Upper Serpulit	*Riasanites*		
UPPER VOLGIAN				Runcton Beds	M. PURBECK — Lulworth Beds; L. PURBECK	Lower Serpulit	Brackish and Brackish-marine		?—
			Lower Spilsby Sandstone			Münder Mergel		TITHONIAN (pars)	
MIDDLE VOLGIAN (pars)				Roxham Beds	Portland Stone		'Portlandian'		

FIGURE 2.2 Suggested correlation of the Jurassic–Cretaceous boundary beds in England, N.W. Germany and Poland (Casey 1973).

seems to pass laterally into normal marine Ryazanian. Such a correlation seems to be broadly supported by other work. Thus Anderson and Hughes (1964) correlated the Middle Purbeck Beds with the German Serpulite and lower part of the German Wealden. The investigations by Norris (1969) on miospores in southern England and by Neale (1967) on ostracods of the type Berriasian have led to results consistent with Casey's views. Casey suggested that the Jurassic part of the Purbeck Beds be called the Lulworth Beds and the Cretaceous part the Durlston Beds. In his most recent published work Casey (1973) puts forward his proposals for correlation of the English Volgian and basal Cretaceous deposits with classic European successions (fig. 2.2) and erects five ammonite zones for the Lower Spilsby Sandstone of Lincolnshire and its Norfolk equivalents. Of particular interest is his suggestion that the Giganteus Zone of the Portland Stone in Dorset correlates with the base of the Lower Spilsby Sandstone.

Because the term Portlandian can be avoided outside the limited area of southern England and northwestern France and because the beds of which it is comprised do not extend to the top of the Jurassic, there is a strong case for dropping it, because stages should have validity over an extensive region. Another difficulty is that its base, either the usually accepted one between the *Pavlovia* zones and the zone of *Zaraiskites albani* or the

TABLE 2.8 *Tentative correlation of the Tithonian, Volgian and English Upper Kimmeridge Clay and Portland Beds*

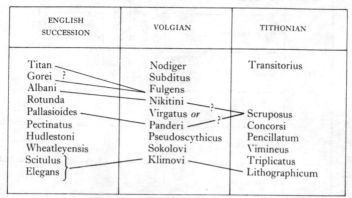

ENGLISH SUCCESSION	VOLGIAN	TITHONIAN
Titan	Nodiger	Transitorius
Gorei ?	Subditus	
Albani	Fulgens	
Rotunda	Nikitini	
Pallasioides	Virgatus *or* ?	Scruposus
Pectinatus	Panderi ?	Concorsi
Hudlestoni	Pseudoscythicus	Pencillatum
Wheatleyensis	Sokolovi	Vimineus
Scitulus ⎤	Klimovi	Triplicatus
Elegans ⎦		Lithographicum

alternative put forward by Casey, between the *Pectinatites* and *Pavlovia* zones, bears no relation to stage boundaries elsewhere in the world. On the other hand, the precise correlation with the Volgian is not yet unequivocally established and the force of tradition is considerable. It will not, however, be used further in this book.

Correlation between the Tithonian and Volgian of Europe and their southern English equivalents is bound to remain tentative and limited (table 2.8). As indicated earlier, *Gravesia* provides the only good link. The Upper part of the Kimmeridge Clay has enough elements in common with the Lower and Middle Volgian (successively species of *Pectinatites* and *Pavlovia*), and the discovery of *Epivirgatites* in the Albani Zone of Dorset allows correlation with the Nikitini Zone of the Volga Basin (Casey 1967). Zeiss (1968) has suggested a correlation between the Scruposus Zone of the Tithonian with the Panderi Zone of the Volgian on the basis of the common occurrence of *Zaraiskites zarjskensis*, but the *Zaraiskites* horizon in England is correlated with the Nikitini Zone. Whichever correlation is preferred, the Transitorius Zone of the Upper Tithonian is likely to represent a greater segment of time than the other zones and further subdivision seems desirable.

ALTERNATIVE METHODS OF STRATIGRAPHIC SUBDIVISION

Although ammonites provide by far the most precise means of correlating and subdividing Jurassic strata they are absent from or rare in many rock successions and so alternative means have to be used, based on the stratigraphic distribution of other fossil groups.

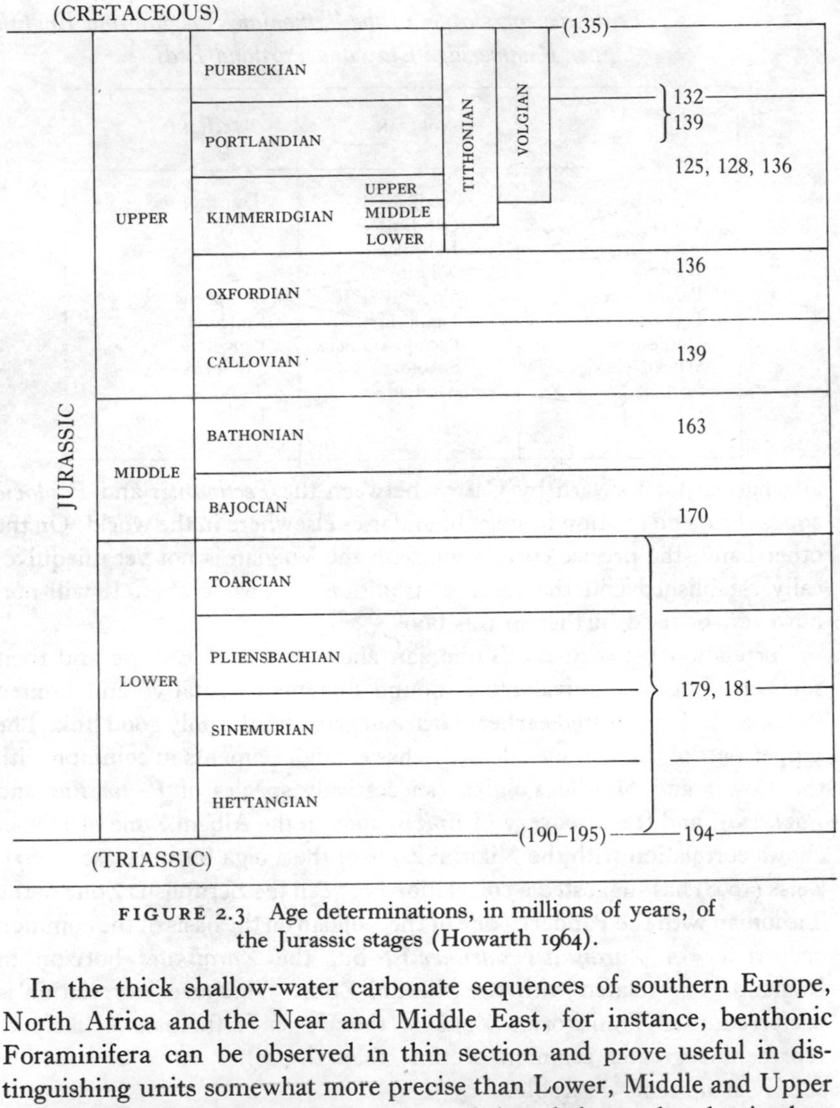

(CRETACEOUS)

JURASSIC							
						(135)	
	UPPER	PURBECKIAN		TITHONIAN	VOLGIAN		} 132
		PORTLANDIAN					} 139
		KIMMERIDGIAN	UPPER				125, 128, 136
			MIDDLE				
			LOWER				
		OXFORDIAN					136
		CALLOVIAN					139
	MIDDLE	BATHONIAN					163
		BAJOCIAN					170
	LOWER	TOARCIAN					
		PLIENSBACHIAN					} 179, 181
		SINEMURIAN					
		HETTANGIAN					
				(190–195)			194

(TRIASSIC)

FIGURE 2.3 Age determinations, in millions of years, of the Jurassic stages (Howarth 1964).

In the thick shallow-water carbonate sequences of southern Europe, North Africa and the Near and Middle East, for instance, benthonic Foraminifera can be observed in thin section and prove useful in distinguishing units somewhat more precise than Lower, Middle and Upper Jurassic (Hottinger 1971). Codiacean and dasycladacean benthonic algae have been used with success in southern Italy (Praturlon 1966). A variety of other groups recognisable in thin section provide quite important stratigraphic indicators in both surface and subsurface exploration. A series of well-illustrated publications offer assistance in this respect, such as those for Aquitaine (Carozzi *et al.* 1972), Yugoslavia (Radoičić 1966), northern Italy (Cita 1965), the western Carpathians (Misik 1966) and Israel (Derin and Reiss 1966).

The marine bivalve genus *Buchia* occurs in huge numbers in some of the latest Jurassic and early Cretaceous deposits of western North America, and have proved to be zonally useful (Jeletsky 1966, Jones *et al.* 1969). *Inoceramus* has likewise been utilised in the Middle Jurassic of the Soviet Arctic.

Planktonic micro- or nano-organisms can also be used, such as coccoliths (Noël 1965, 1972), the non-calcareous acritarchs and dinoflagellates (Sarjeant 1964) and, in the Upper Tithonian pelagic limestones, calpionellids (Remane 1964).

In non-marine strata, pollen and spores are available (Couper 1958) and certain bivalves have proved their utility in Soviet deposits (Martinson 1964); zonation of the Purbeck Beds can be achieved using ostracods (Anderson and Hughes 1964).

RADIOMETRIC AGE DETERMINATIONS

In his review of Jurassic radiometric age determinations, as a contribution to the *Phanerozoic Time Scale* published by the Geological Society of London, Howarth (1964) concluded that the period had lasted between 55 and 60 million years, from 190–195 to 135 million years ago (fig. 2.3). More recently, age determinations of post-Triassic plutons in western North America suggest that the lower limit of the period should perhaps be extended to slightly over 200 million years (Harland 1971). Since there are some 65 ammonite zones established for the northwest European Jurassic, the average time value for a zone is about one million years.

3 Arenaceous, argillaceous and ferruginous facies in northern Europe

Though arenaceous, argillaceous and ferruginous deposits occur together in intimate association, it will be convenient to consider them separately in some detail before discussing some of the broader implications of the facies relationships. Generalisations are not easily made about the arenaceous deposits because they are subordinate in volume to the argillaceous deposits and have not attracted a great deal of attention. Much more than with the other facies, we are restricted to citing and commenting on the results of the small number of detailed analyses that have been undertaken, while resisting what would be a premature attempt to draw conclusions of general significance.

ARENACEOUS ROCKS

It is one of the curiosities of sedimentology that study of the obvious sedimentary structures of sandstones was long neglected in favour of more esoteric approaches, firstly the analysis of heavy minerals and then granulometry or grain size analysis. Both types of study are now out of fashion, to a considerable extent because disillusionment set in as their limitations as a guide to environment and stratigraphic correlation became progressively more apparent. Nevertheless it must still be acknowledged that useful information on provenance can often be obtained from heavy mineral analysis, provided that full allowance is made for selective destruction of certain minerals by weathering in the source area and diagenesis within the sandstone deposits. Indeed, heavy minerals and pebbles offer about the only means of identifying source areas, as opposed to directions of local sediment transport. This can be well exemplified from studies made on Jurassic sandstones in northwest Europe.

Many years ago Boswell (1924) undertook pioneering work on the Toarcian and Aalenian sands of southwest England which, judging from their eastward and northeastward passage into clays, were derived from somewhere to the west or southwest of the present outcrop. He recognised a suite of heavy minerals including garnet, kyanite, staurolite, sillimanite, epidote and sphene, which he considered must have had a source in regionally metamorphosed rocks such as currently outcrop in western Brittany. A southwesterly provenance was therefore inferred. Subsequent analysis of Lower Cretaceous sandstones in southern England confirms the inference of an extensive source area of metamorphic rocks, occupying the site of the Western Approaches, which persisted into the Cretaceous. Thus the heavy mineral suite of the Upper Kimmeridgian and Portland beds in southern England is almost identical to that of the Upper Lias Sands (Neaverson 1925), and Allen (1969) inferred a westerly metamorphic source area for some of the constituents of the southern English Wealden sands.

The work of Smithson (1942) on Middle Jurassic sandstones in northeast England (the Cleveland Hills of east Yorkshire) gave an indication of the complicating effect of diagenesis. He found that at given horizons lateral variations of the heavy mineral assemblages were considerable. The heavy minerals were in general most impoverished in the northeast of the region studied, except for authigenic minerals such as brookite. This Smithson attributed to diagenetic solution long subsequent to deposition.

From the grain size distribution of stable minerals both a Pennine landmass to the west and a narrow North Sea landmass to the east were inferred, with northeast Yorkshire being an intermediate area of delta and swamp, the deposits of which were first analysed sedimentologically by Black (1929). The evidence of sediment grain size, current directions and marine phases converged to suggest a sediment source area to the north-northwest, and the heavy minerals tended to confirm this by indicating the Scottish Caledonian mountain belt as the likeliest source for a number of metamorphic minerals. Smithson's evidence for local Pennine and North Sea landmasses does not seem very cogent, and his dubiously justifiable inferences on this subject could be held to illustrate the dangers of extrapolating a regional picture too boldly from geographically very limited outcrops.

It has become increasingly apparent over the years that mere presence–absence data concerning a few species of heavy minerals is almost totally uninformative, and that relative abundance may be far more significant, though it should be noted that the presence of a given mineral in reasonable amounts always needs accounting for, whereas its absence may be due to a variety of factors. Smithson made a beginning in this direction and Hudson (1964), in his thorough study of the petrology of sandstones of the Great Estuarine Series (Bathonian) of western Scotland, found it profitable to use quantitative, though not strictly statistical, methods. He accordingly confined himself to considering the major variations in heavy mineral content.

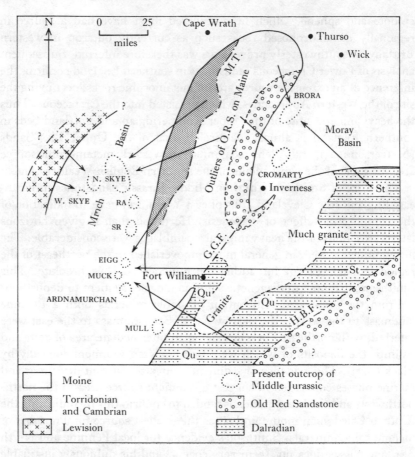

FIGURE 3.1 Speculative palaeogeographic reconstruction of northern Scotland for the Bathonian. Arrows show postulated net movement of sediment. M.T., Moine Thrust; H.B.F., Highland Boundary Fault; G.G.F., Great Glen Fault; St, Staurolite schist; Qu. quartzite in Dalradian; SR, Strathaird; RA, Raasay; ————, generalised geological boundary; – · – ·, major fault (Hudson 1964).

Hudson was able to take account of and climinate certain factors in his analysis of sediment provenance. Kyanite and sphene, for instance, were found to be almost completely restricted to cemented beds, which suggests differential loss of these minerals by diagenesis. The almost total absence of hornblende was attributed to weathering in the source region. A grain size effect was also discernible: tourmaline tends to be concentrated in the coarser-grained beds, zircon in the finer.

Two source areas were distinguished for the Hebridean deposits (fig. 3.1). The Precambrian Moine and Dalradian rocks of the Scottish mainland seem to be the obvious source for the abundant fresh garnet and also staurolite of the main outcrop, also suggested from pebbles of Moine-type schist. Rounded zircons, rutile and iron ore are thought to be derived from a sedimentary source, probably the late Precambrian Torridonian deposits. Chert pebbles are presumed to come from the Cambro-Ordovician Durness carbonate sequence. The deposits of northwest Skye have a distinct suite of minerals characterised by an abundance of epidote, together with garnet, apatite, and sphene. These are thought to have their source in basic orthogneisses of the Lewisian basement complex.

Heavy mineral analysis forms the major part of Larsen's (1966) study of the terrigenous clastic Rhaetian to Lower Cretaceous sediments of the Danish Embayment, which are comprehensively known only from borehole cores. As regards the factors which can bias environmental interpretation, Larsen sometimes came to different conclusions from Hudson. Thus zircon was held to be concentrated in the coarse-grained sediments, in contrast to tourmaline, which is more abundant in the finer sediments. Hornblende seems to have been little affected either by source-area weathering or by diagenesis, a rather surprising conclusion. In general, differential removal of certain mineral species was not considered an important modifying factor.

Larsen inferred the existence in the north of his study area of a province rich in epidote, with accompanying kyanite and staurolite, which signified an ancient metamorphic complex in Fennoscandia as the source. The Rhaeto-Liassic deposits are different from the rest in their richness in garnet and feldspar, also thought to be derived directly from metamorphic rocks but probably from a region further south than for the epidotic suite. Some minerals have evidently been reworked from older sediments exposed on the Ringkøbing–Fyn High, which flanks the Danish Embayment on its southern side.

Sandstone sedimentology at the present day is very largely concerned with the interpretation of sedimentary structures. Their great value in elucidating Jurassic depositional environments is illustrated by several recent publications.

Häntzschel and Reineck (1968) undertook a detailed study of a sequence of sandstones, siltstones and shales in the Hettangian of northwest Germany (Lower Saxony). Structures in the sandstones include horizontal and lenticular bedding with flaser texture in the rocks of mixed facies, oscillation and interference ripples, chevron marks and groove casts. At the base of the sandstones small channels have eroded into the underlying shales. There is no evidence of lateral displacement of these channels and hence a tidal flat origin is ruled out; they were evidently formed by

27

short-lived minor water courses. A wide variety of trace fossils is found on the sandstone soles, including *Chondrites, Asteriacites, Gyrochorte, Teichichnus, Thalassinoides, Rhizocorallium, Curvolithus* and *Neonereites*. This is an assemblage characteristic of many Lower and Middle Jurassic assemblages in Germany. Together with the evidence of inorganic structures, they signify a shallow subtidal regime, showing no evidence of tidal flat emergence. A close comparison can be made with sediments at present being deposited at between 6 and 40 m depth in the German Bay, off the North Sea. The thin sandstone intercalations are probably the result of periodic storms transferring sand to slightly deeper, quieter water (Reineck and Singh 1972).

Similarly, a careful study by Knox (1973) of the Eller Beck formation, a Bajocian marine intercalation within the Middle Jurassic deltaic deposits of Yorkshire, consisting of ironstone followed by a coarsening-up sequence of shale and sandstone, suggested little to no tidal activity. The deposits show only modest geographic variations and seem to have been laid down in extremely shallow water, but exhibit no tidal channelling; they were very sensitive to minor tectonic movements. Deposits of this type are quite widespread in northern Europe.

In contrast to these other workers, Sellwood (1972a) found what he considered to be conclusive evidence of tidal flat sedimentation in Sinemurian deposits on the island of Bornholm, in the Baltic Sea. He recognised a regressive sequence of deposits, somewhat analogous in those occurring in the Wash of eastern England, which had been interrupted by minor tilting and erosion.

At the base is a sandy unit exhibiting planar, tabular and herring-bone cross bedding, together with clay drapes. This unit seems to signify a shallow subtidal environment in which pulsating tidal currents were interrupted by phases of stillstand, when mud was enabled to settle and hence lead to the formation of clay drapes. Overlying this unit, in succession, are firstly a series of beds with more sand than clay, interpreted as signifying a low tidal flat environment, and then a series of clay deposits with sand lenticles, exhibiting wavy and flaser bedding, interpreted as a high tidal flat unit. The sequence is capped by coals, signifying emergence. Within the presumed tidal flat sequence are a series of sand-filled tidal channels, together with some flat-bedded sheet sands probably introduced by wind tides of hurricanes.

Another attempt to relate arenaceous lithologies and structures to environments has been made by Davies (1969) for the diachronous Toarcian sands of southwest England (fig. 3.2). Major facies changes at a given horizon (usually an ammonite subzone) can be recognized (fig. 3.3).

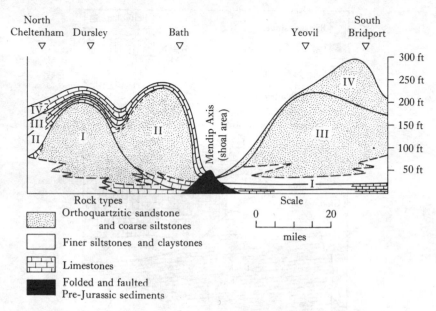

FIGURE 3.2 Stratigraphic cross-section of Toarcian sediments in southwest England. I, Cotteswold Sands; II, Midford Sands; III, Yeovil Sands; IV, Bridport Sands; — time line; --- facies boundary (Davies 1969).

Heavily bioturbated siltstone in the south is related to a fore bar; fine-grained laminated and ripple drifted sandstone immediately to the north was interpreted as a bar deposit, cut by a presumed tidal channel containing biosparites. More bioturbated siltstone north of this corresponds to a back bar. Only the Ham Hill Stone biosparite, a local lens-shaped deposit up to 25 m thick, has large-scale cross bedding, which fines upward into small-scale ripples. This deposit, according to Davies, occupies a tidal channel cut into the bar sands.

As for the provenance of the sands, they must have come from the southwest, as pointed out earlier in this chapter. Davies believed the sediments to have been carried by longshore drift adjacent to a land lying to the west. Cross bedding measurements, however, indicate, a dominant current flow from the northeast, which is a somewhat unexpected finding.

It is open to question whether Davies had enough data to produce such an ambitious environmental model. The outcrop is only a narrow one and runs roughly from north to south, exactly in the direction of the principal facies changes. One wonders, moreover, why the bar deposits do not show large-scale cross bedding, and why the still-enigmatic Ham Hill Stone of Somerset is composed of shell hash rather than sand.

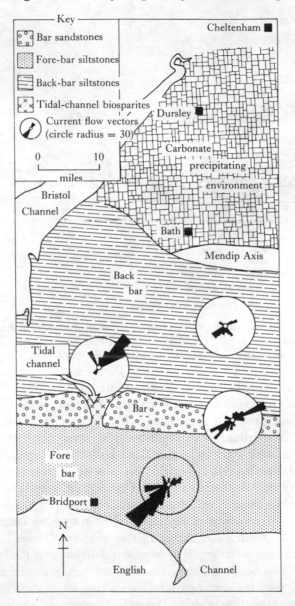

FIGURE 3.3 Spatial arrangement of depositional environments and sediment dispersal directions developed in Upper Toarcian (*D. moorei* Subzone) times (Davies 1969).

Davies has in another paper (1967) discussed the origin of the regular small-scale alternations of calcareous cemented and friable sands which characterise the deposits in question and which are seen excellently displayed in the coastal cllffs of Bridport, Dorset. He favoured an origin in terms of primary differences in sedimentation enhanced by diagenetic migrations of $CaCO_3$, rather than secondary, post-depositional unmixing of $CaCO_3$ from an originally homogeneous series of deposits. The latter hypothesis, it was argued, fails to explain (i) the observed relationship of calcareous horizons to an original bottom configuration, (ii) the bifurcation of such horizons into distinct beds separated by friable sandstone, and (iii) the sharpness of the change in $CaCO_3$ content from bed to bed.

A personal examination of the beds in question suggests that point (i) is only approximately true and that, very generally, there is no obvious relationship to be observed between the calcareous–friable sand alternations and the scour and bioturbation features ubiquitous in the Bridport Sands. Furthermore, points (ii) and (iii) are unconvincing. Why should not the change in $CaCO_3$ content be sharp? We know too little as yet of diagenetic unmixing mechanisms.

ARGILLACEOUS ROCKS

The argillaceous rocks may variously be termed shales, mudstones or clays depending on their fissility and plasticity, and more fully described by such adjectives as sandy, calcareous and pyritic. Frequently 'shale' has been used in a broad sense as a conveniently short way of grouping together a wide variety of argillaceous rock types. There seems to be no great objection to this, provided a fuller description is also given in terms of structure and composition. The Jurassic 'shales' of northern Europe may be conveniently discussed in terms of their clay mineralogy and content of detrital minerals, calcite, siderite, organic matter and fossils.

Clay minerals

By far the most important clay mineral is illite, which usually accounts for the bulk of the rock. Kaolinite is the most common subordinate clay mineral, normally less than 5 to 10 per cent of the total (Hallam and Sellwood 1968) but exceptionally exceeding 60 per cent, as in Lower Lias non-marine facies in northeast Scotland (Sellwood 1972b). The ratio of illite to kaolinite appears indeed to be a useful environmental parameter, being lowest in non-marine deltaic facies, intermediate in inshore marine regions where there is a strong admixture of silt and sand signifying the proximity of river deltas, and lowest in offshore marine regions where the clays contain less detrital quartz and more $CaCO_3$ (Hallam 1967 *b*; Sellwood 1972 *b*). This pattern matches that found in marine clays at the present day where kaolinite is concentrated in the coastal regions in low latitudes (Griffin *et al.* 1968).

Of the other clay minerals, chlorite appears to be virtually absent and mixed-layer illite–montmorillonite occurs usually in no more than trace quantities. Of great interest is the local occurrence in southern England of pure montmorillonite.

Rocks known as Fuller's Earths have been exploited economically in southern England for a considerable period of time, initially for fulling wool but in recent years for a variety of industrial purposes, all of which depend on the peculiar chemical and physical properties of the dominant clay mineral, montmorillonite. Two such deposits, each a metre or two thick, occur in the English Mesozoic. One is Aptian in age and therefore not strictly the concern of this book; the other, in Somerset, is Bathonian. Both deposits consist of pure or almost pure montmorillonite, but with small quantities of authigenic feldspars, zeolites and cristobalite–tridymite and rare detrital biotite (Hallam and Sellwood 1968). Montmorillonite associated with some of these other minerals has also been recorded from certain Oxfordian deposits in southern England (Brown *et al.* 1969). Montmorillonite is also recorded from Upper Oxfordian clays in Dorset (Brookfield 1973 *a*).

Such unusual deposits call for a special explanation, which is that they are true bentonites, i.e. they have formed from the chemical breakdown of fine volcanic ash which has dropped into sea water after eruption (Hallam and Sellwood 1968). The most direct evidence in support of a volcanic interpretation would be the presence of glass shards, but these have not been convincingly demonstrated despite older claims. Their absence is not critical, however, because such structures are not normally preserved in the finer ash deposits for long periods of geological time, and indeed have not been located in the classic Cretaceous bentonites of Wyoming despite an intensive search. In all the other characteristics they closely resemble such classic bentonites. If montmorillonite had been formed in soils on the land it is difficult to conceive how this mineral would not have been considerably diluted by the ubiquitous illite and kaolinite as it was transported into the sea. The elaborate non-volcanic explanation adopted by Brown *et al.* (1969) seems to us totally unconvincing, as argued in our reply to their article (Hallam and Sellwood 1970).

As far as the location of the volcanoes, data are still sparse, but evidence has recently been cited which suggests the likelihood of considerable Cretaceous volcanic activity at no great distance from southern England

(Cowperthwaite *et al.* 1972). With regard to the Jurassic, recent exploration for petroleum in the northern North Sea has established the existence locally of thick basaltic rocks, probably of Bathonian age, associated apparently with an incipient phase of tension in the North Sea graben. The North Sea therefore appears to be the likeliest source of the Bathonian Fuller's Earth, together with more limited amounts of montmorillonite in the Upper Oxfordian deposits, which could have been reworked during a marine transgression.

Detrital quartz and Feldspar

These minerals are present as angular fragments of fine to coarse silt grade in all the shales, but only in minute quantities, considerably less than 2 per cent by volume, in the smoother-textured varieties. In the proximity of contemporary river deltas, as in most of the European Jurassic north of latitude 53°N, the detrital content of quartz and subordinate feldspar increases sufficiently to give the rocks a rough texture. Thin section analysis reveals that the grain size ranges up to fine sand grade. Muscovite flakes also occur in variable quantities: in the Hebridean Jurassic they are strikingly abundant, presumably because of erosion of the Moine Schists on the Scottish mainland.

Calcite and siderite

Where the argillaceous rocks contain more than about 25–30 per cent of calcite, as they frequently do in association with limestone beds, they acquire a conchoidal fracture and increased hardness and may be described as marls. For the more usual 'shales' the gross content of $CaCO_3$ is normally less than about 10 per cent but the calcite is normally concentrated as layers of ellipsoidal nodules or concretions, sometimes exhibiting septarian structure, containing over 80 per cent $CaCO_3$. Their origin being obviously diagenetic, they need not concern us further here, but it should be noted that the question of whether the concretionary horizons develop as a result of primary sedimentational differences is still not adequately resolved.

Siderite forms nodules in a manner strictly analogous to calcite. There seems to be a definite association between the presence of siderite and detrital silt and sand. This was investigated quantitatively by examining samples of carbonate nodules taken from successive ammonite zones in the Lias of the Yorkshire coast (Hallam 1967a). The ratio of calcite to siderite in each nodule was determined by X-ray diffractometry. By using the potassium ferricyanide and alizarin red S staining techniques on thin sections it can be demonstrated that, in those rocks containing both

33

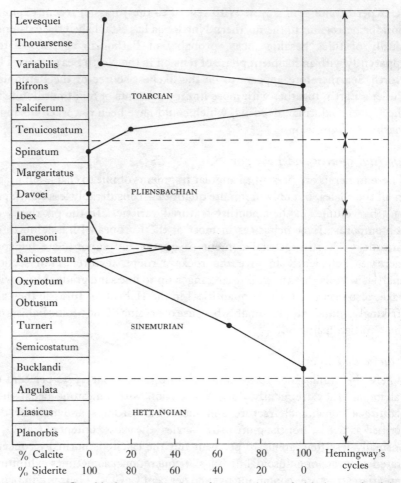

FIGURE 3.4 Graphical representation of composition of siderite- and calcite-bearing nodules in different zones of the Liassic of Yorkshire. 1, 2 and 3 correspond to major sedimentary cycles of Hemingway (1951) (Hallam 1967 a).

siderite and calcite, the siderite crystallised first in small micritic patches about 0.5 mm in diameter. The matrix later became cemented by ferroan calcite microsparite, presumably when local supplies of dissolved iron became exhausted.

The graphical plot (fig. 3.4) shows that striking changes in the proportion of calcite to siderite take place up the Liassic succession. These correlate in a general way with the change from smooth-textured shales to silty–sandy shales and silty mudstones and back again, as expressed for instance by several major sedimentary cycles distinguished by Hemingway

(1951). These can be related to the progressive advance and retreat of a river delta at no great distance from the area of deposition (Hallam 1966). Thus the proportions of siderite and calcite in early diagenetic concretions appears to provide a fairly sensitive index of environmental change.

Sellwood (1971) has addressed himself to the problem of why siderite nodules have formed in the Lias of Yorkshire rather than pyrite, which on chemical grounds would seem more likely (e.g. Curtis and Spears 1968). According to his hypothesis, iron is brought into the depositional area as particles of oxide attached to clay mineral flakes and is subsequently mobilised in anaerobic conditions within fine-grained sediment. In the ferrous form it migrates to the sediment–water interface and is precipitated as ferric iron in highly bioturbated beds whose faunas indicate oxidising conditions. Renewed sedimentation leads to the resumption of reducing conditions and the further mobilisation of ferrous iron. No sulphide ions produced by anaerobic bacteria are present because free access of sulphate from the sea water is prevented by a seal of accreting clay. Hence siderite is precipitated rather than pyrite.

Organic matter

Most of the shales contain only negligible quantities of organic matter, in finely particulate form, but two types must be excepted. Firstly, in non-marine coal measure facies, as for instance in the Middle Jurassic of Yorkshire, certain shales are richly carbonaceous, with abundant fragments of macroscopic plants. Thin bands of carbonaceous shales, with woody fragments, mottled by *Chondrites* and alternating cyclically with paler-coloured shales, have also been recognised by Mr R. M. Sykes in the Callovian and Lower Oxfordian of northern Skye; they are thought to represent periodic incursions of material from the nearby land.

The second category are marine and can be described as laminated bituminous shales; in weathered outcrops the wafer-thin laminae disaggregate, hence their description as paper shales. The origin and environmental significance of this latter group of rocks is of considerable interest and hence will be discussed at some length.

Microscopic examination reveals that these shales contain closely-spaced bituminous laminae, with the clay-organic rhythmic couplets averaging about 20–30 microns in thickness. The bituminous matter is largely microscopically structureless kerogen, red– or yellow–brown and partly translucent in thin section, but there are also small quantities of recognisable plant debris together with spores, pollen and microplankton (e.g. Wall 1965). The total content of organic carbon may range, rarely, up to 15 per cent, but is normally less than 7 per cent (Bitterli 1963a). Early diagenetic pyrite, as microscopic cubes or framboidal aggregates, is normally quite common. Besides the usual clay minerals and detrital quartz, fine grained calcite is present, usually as a small percentage, but increasing to over 80 per cent in scattered calcareous concretions. Examination with the scanning electron microscope has shown that coccoliths are present in certain southern German Liassic deposits, the Posidonienschiefer (Müller and Blaschke 1969). Benthonic fossils are sparse or absent; those that occur tend to be small and thin-shelled.

Such deposits, though much less common than other types of shale, are widely distributed both geographically and stratigraphically. Thin horizons intercalated in other types of deposits occur in the Hettangian of both England and Germany, and locally extend up into the Lower Sinemurian, as in southwest England (Hallam 1960). Pliensbachian (Davoei and Margaritatus zones) deposits occur in the thicker basinal facies of the eastern Cantabrian mountains, northern Spain (Dahm 1966), and west-central Portugal (Hallam 1971 c). A Lower Toarcian (mainly Falciferum Zone) horizon is especially widespread, and can be traced all the way from central and northern England to southern France (Aveyron) and West Germany. Higher in the southern English Jurassic, thick bituminous shales occur in both the Middle and Upper Callovian (Lower Oxford Clay) and the Kimmeridgian. The extent to which these shales are finely laminated has not, however, been adequately determined. It appears indeed that the southern English Lower Oxford Clay is not composed of typical laminated shales, and they contain a rich benthos. On the other hand, correlative beds in northern Skye do appear to be laminated and almost devoid of benthos.

The widespread Lower Toarcian horizon includes both the celebrated Posidonienschiefer of Baden–Württemberg, Germany, and the Jet Rock of Yorkshire, England. The Posidonienschiefer are so named because of the abundant occurrence at certain horizons of the bivalve *Bositra* (= *Posidonia* auct.) which is commonly considered to have had a planktonic habitat (e.g. Jefferies and Minton 1965). Likewise, the abundant *Inoceramus* of the Posidonienschiefer and Jet Rock might have fallen to the sea bed after death having spent their adult life at the surface attached by byssal threads to driftwood or floating seaweed. This is clearly suggested by the discovery of specimens of driftwood associated with *Inoceramus* (Hauff 1953). Hauff also illustrated superb specimens of crinoids from the Posidonienschiefer of Holzmaden which were clearly attached during life to driftwood. A detailed reconstruction of the functional morphology of the pseudoplanktonic *Seirocrinus* has been made by Seilacher *et al.* (1968).

While undoubted benthonic fossils are either absent or extremely sparse, nektonic vertebrates such as ichthyosaurs and fish are beautifully preserved as complete specimens, often with a carbonaceous film signifying the outline of the skin. These and other well-preserved organisms, such as crocodiles and crustaceans, from the Holzmaden deposits have been excellently illustrated by Hauff (1953). Such fine preservation is not peculiar to the Posidonienschiefer but merely reflects the fact that these deposits have been worked over in detail by quarrymen over many years. The stratigraphic term 'Fish and Insect Beds' for correlative deposits in the English Midlands is illuminating in this respect. Many of the finest vertebrates from the Lower Lias of Lyme Regis, another celebrated fossil collecting locality, came from the bituminous facies of the Sinemurian.

The Swabian Posidonienschiefer also contain a series of early diagenetic concretionary limestone horizons, either as regular bands or small ellipsoidal nodules; their origin is discussed in detail by Einsele and Mosebach (1955). Concretions are also a notable feature of the Yorkshire Jet Rock. Some are ellipsoidal to spheroidal and possess pyritic coatings: others are huge ellipsoids (the 'Whalestones') with evidence of a more than one episode of $CaCO_3$ segregation: a third category, the so-called 'Pseudo-vertebrae', are especially curious and contain both grey and brown micritic calcite with evidence of colloidal banding, pseudobrecciation and physical disruption of laminae (Hallam 1962).

Detailed accounts of the Jet Rock are given by Hemingway (1958, 1974) and Hallam (1967c) and the geochemistry of this and other Yorkshire Toarcian shales discussed by Gad et al. (1969). The term jet signifies a shiny, black lignitic material. This has been produced from the alteration of scattered fragments of driftwood by the permeation and partial replacement of woody cells by fairly reactive colloidal or fluid organic compounds, derived from the decay of organic matter either in the associated shales or from other parts of the driftwood. It is by no means confined to the Yorkshire Jet Rock but occurs widely in shaly deposits of various types, and often exhibits a significant enrichment in germanium and vanadium as a result of selective adsorption by the colloidal organic matter during or shortly after deposition (Hallam and Payne 1958). In more calcareous deposits the driftwood preserves its woody character owing to the early diagenetic crystallisation of calcite within the plant cells. The woody lignite fragments are normally much larger in size than the jet, signifying shrinkage of the latter from its original state.

The conditions of deposition of laminated bituminous shales such as the Posidonienschiefer have been frequently discussed (e.g. Bitterli 1963b) and general agreement has been reached on one point: that the bottom waters must have been stagnant, i.e. anaerobic. This explains at once the absence of burrowing benthos, which would have disrupted the laminae, the presence of abundant pyrite, the lack of oxidative destruction of bituminous matter and the excellent preservation of vertebrates, which have not suffered attack by scavengers. Certain qualifying observations need to be made, however. The presence of pyrite certainly signifies anaerobic conditions resulting from the activities of sulphate-reducing bacteria, but such conditions commonly obtain within a wide variety of fine-grained sediments which were deposited in oxygenic waters. Since pyrite forms diagenetically from the monosulphide hydrotroilite its formation need have no direct bearing on the condition of the bottom waters.

The presence of organic laminae is a much safer criterion of oxygen deficiency on the sea bed. Even here some caution should be exercised, bearing in mind the results of research on possible modern analogues. Both Hülsemann and Emery (1961) for the Santa Barbara Basin off southern California, and Calvert (1964) for the Gulf of California, have described sediments with organic laminae which are being laid down in bottom waters with small amounts of oxygen. This is evidently sufficient to support a sparse fauna of soft-bodied dwellers on the sediment surface but insufficient to support burrowers which would destroy the lamination (see also Rhoads and Morse 1971). Such observations may well throw some light on the sparseness of apparently benthonic epifaunal bivalves in the Jet Rock (Hallam 1967c) and comparable deposits. Presumably the oxygen content of the bottom waters fluctuated in time from low to zero. At times when oxygen was present in small quantities, burrowed horizons formed such as the *Chondrites*-bearing Seegrasschiefer of Baden–Württemberg.

With regard to the origin of the rhythmic couplets of clay and organic matter, further analogies with Recent sediments, such as those being deposited in parts of the Black, Adriatic and Clyde Seas, suggests rather strongly that they are annual and hence are varves, the organic layers corresponding to the late spring or early summer bloom of phytoplankton (Hallam 1960, 1967c). Rates of sedimentation of the compacted shale of the order of 0.3 mm year^{-1} are indicated.

Interpretation of the more general problem of the depth of deposition and the regional palaeogeographic framework of these interesting deposits is more controversial and will be postponed until the end of the chapter.

Fossils

In so far as fossils are important components of the argillaceous facies their mode of preservation should be noted. Calcitic fossils such as brachiopods, crinoids and ostreiform and pectinid bivalves are normally preserved in their original state. Aragonitic fossils are less well preserved but by no means uncommon in the form of nacreous or chalky shell material, which

X-ray diffractometry reveals to be either pure aragonite or substantially aragonite with slight alteration to calcite (there seems to be no correlation with the degree of alteration to calcite and the breakdown of the nacreous shell structure). Ammonites and other molluscs with this mode of preservation are abundant, for instance, in the Oxford and Kimmeridge Clays of England, and even occur in deposits as old as the Lower Lias. Many formerly aragonitic shells, for instance deep-burrowing bivalves such as *Pholadomya* and *Pleuromya*, are, however, preserved only as moulds. In marked contrast, aragonite is rarely preserved in Jurassic sandstones or limestones, where it has almost always been dissolved and the cavities either left vacant or filled with secondary calcite.

Pyritic preservation is also common, having occurred by the occupation in early diagenesis of internal cavities not yet filled by sediment; the result is pyritic internal moulds of formerly aragonitic shells. The different types of pyritisation of fossils in the Oxford Clay of Buckinghamshire have been dealt with in detail by Hudson and Palframan (1969), who have shown, for instance, how the nuclei of ammonites have been selectively preserved in the shales while the body chamber and outer part of the phragmocone has been replaced by calcite only in the limestones. Unless this differential preservational factor is appreciated, misleading interpretations could be made of the considerably different sizes of the preserved ammonites in the two lithologies. It should also be noted that in parts of the Lower Oxford Clay pyrite has actually replaced aragonite rather than merely infilling cavities. Quite why certain horizons, such as the Raricostatum Zone of the British Sinemurian, should exhibit widespread pyritisation of fossils has not yet been explained.

Many shales which alternate with sandy or calcareous beds show a conspicuous mottling which signifies bioturbation by burrowing benthos. A number of trace fossil ichnogenera can be recognised which are sometimes also preserved as calcitic or sideritic concretions, indicating the migration of the relevant ion-bearing fluids in early diagenesis, presumably controlled by porosity differences within the burrowed sediment.

IRONSTONES

Although volumetrically subordinate, Jurassic ironstones in Europe have attracted a great deal of attention over many years, both because of their economic importance and the intriguing probems of environment of deposition that they have posed.

Characteristically the ironstones, bluish green when fresh and gingerbread brown when weathered, consists of 'chamosite' ooliths, sometimes oxidised to goethite or, more rarely, haematite, in a matrix of siderite microsparite; other minerals include calcite, either as shells or sparry matrix, and, less commonly, pyrite, collophane and opal. Such beds may be intercalated with beds of siderite microsparite (or 'mudstone'), oolitic goethite with sparry calcite matrix, or chamositic, pyritic shale. The coarser-grained beds may exhibit cross bedding and scour surfaces, with scattered horizons of interformational conglomerate, all signifying conditions of high water energy. Marine fossils are usually common, with bivalves, crinoids and brachiopods dominant; ammonites may be common in those ironstones with a mudstone matrix. Bioturbation is also common, and a number of trace fossils have been recognised and described (e.g. Hallam 1963a, Farrow 1966).

Where the stratigraphic control is adequate, the ironstones are seen generally to be condensed compared with their lateral correlatives in other facies. They are usually only a few metres thick and rarely exceed 15 m. Geographically a given ironstone will rarely extend over more than a few hundred km² (usually less than one hundred) before passing into less ferruginous rock with a higher proportion of terrigenous clastic material. By and large the rock proves exploitable as an ironstone if the total iron content exceeds 20 per cent; some of the best ores contain up to about 40 per cent.

TABLE 3.1   *The distribution of oolitic ironstones in Britain*

Ironstone	Stage	Zone	Locality
Ardnish	SINEMURIAN	Semicostatum	Isle of Skye
Frodingham		Semicostatum–Obtusum	Lincolnshire
Cleveland	PLIENSBACHIAN	Spinatum	Yorkshire
Marlstone			Midlands
Raasay	TOARCIAN	Bifrons	Isle of Raasay
Rosedale		Levesquei	Yorkshire
Northampton Sands	AALENIAN	Murchisoni	Midlands
Dogger			Yorkshire
Westbury	OXFORDIAN	Pseudocordata	SW England
Abbotsbury	KIMMERIDGIAN	Cymodoce	Dorset

The ironstones occur widely in northern and central Europe and examples occur in almost every stage, but they are thicker and more widespread in the Lower and Middle than in the Upper Jurassic. In Great Britain many such ironstones occur (table 3.1) but the only ones of real importance economically today are the Northampton Sand Ironstone, whose petrology is described in exemplary fashion by Taylor (1949), and the Frodingham Ironstone (Hallam 1963a). The Cleveland Ironstone of Yorkshire (Dunham 1960, Chowns 1966, Catt *et al.* 1971) was formerly of great importance for the steel industry of Middlesborough but the ore field is now defunct.

In northeastern France the celebrated minette ores of Lorraine have been the basis of a flourishing steel industry but are now declining in import-ance. The ores occur in two regions which are structurally, rather than stratigraphically, isolated from each other – the more important Briery Basin in the north and the Nancy Basin in the south. In age they are mainly Upper Toarcian but extend locally into the Aalenian. The various beds are termed couches rouges, noires, vertes or brunes according to the dominant mineralogy, i.e. whether the beds are haematitic, pyritic, chamositic or goethitic. It should be noted that French mineralogists still use the term 'chlorite' rather than chamosite for the dominant iron silicate mineral (Bubenicek 1961). This is probably justifiable on the grounds that thur-ingite, another member of the chlorite family, is also present in some deposits (R. M. Weinberg, personal communication).

A number of small ironstone fields, now almost totally defunct, also occur in northwest Germany (Bottke *et al.* 1969). Lower Sinemurian and Lower Pliensbachian ironstones occur in the foreland of the Harz and the Rheinische Schiefergebirge. Upper Aalenian ores are known from bore-holes, and ores of Bajocian, Bathonian and Callovian age are rather frequent in the Gifhorn Trough and near the north coast. Oxfordian ironstones, the only ones still exploited, occur in the Wesergebirge, the Gifhorn Trough and the northern border of the Harz Mountains. A few Aalenian ironstones also occur in southwest Germany.

The principal problems concerning the origin of these ironstones, which, though showing individual differences, are clearly members of a closely-knit family, involve the environmental conditions which control the development of these unusual beds in space and time, and, more specifically, the formation of chamosite. As is well known, argument by actualistic analogy is ruled out because of the lack of closely comparable Holocene deposits.

The orthodox views of minette ore formation assume derivation of iron by continental weathering, from topographically subdued land with a thick vegetation cover in a warm humid climate. A pre-concentration of iron might be achieved under these conditions by the formation of laterite, and iron can persist in solution in the ferrous stage in sluggish rivers con-taining an abundance of protective organic colloids (e.g. Taylor 1949, Strakhov 1967). A radically different hypothesis has been proposed by

Borchert (1960) who believed that the conventional interpretation has several insuperable difficulties and argued that certain deposits on the sea floor, possibly far from land, themselves acted as sources of iron.

Borchert envisaged three depth zones in the sea, with distinctive chemical characteristics and bottom deposits. The deepest is the hydrogen sulphide zone, characterised by anaerobic bottom waters and sediments consisting of pyritic bituminous mud. The shallowest is the oxygen zone, with a rich bottom fauna, where goethite ooliths form in agitated water. Between these is the carbonic acid zone, rich in CO_2, where the bottom deposits consist of sideritic muds; iron is supposed to be mobilised from the deposits of this zone. Part of the iron is carried by marine currents into shallower water where it contributes to the formation of goethite. Part is carried to the H_2S zone where it encounters silica dissolved in alkaline water liberated from the bituminous mud: chamosite forms as a result.

A major difficulty of this interpretation is that the depth zones are hypothetical and implausible in terms of the modern marine environment. There is no evidence, for example, of a widespread CO_2 zone, and H_2S forms only in certain limited areas of restricted circulation. Another problem is that some of the chemical reactions are dubious, for instance those involving the supposed liberation of silica from mildly alkaline water. Apart from these objections there are ample grounds for rejecting Borchert's hypothesis for the Jurassic ironstones, if one takes into account their lateral and temporal facies relationships (Hallam 1966).

If Borchert were correct, there should be a definite association of bituminous shales and ironstones at the same stratigraphic horizon. On the contrary, there is by and large a negative correlation in occurrence, times of widespread bituminous shale formation being marked by a disappearance of ironstones and vice versa. More significant is the association between ironstones and detrital quartz, in the form of thin beds of fine sandstone and thicker beds of silty or sandy shales with sideritic nodules. These facts, together with the relative abundance of kaolinite and driftwood compared with the less sandy and more calcareous facies, lend support to the conventional view of iron derivation, with ferruginous deposits forming close to river deltas (fig. 3.5). The fairly recent discovery of chamosite in deposits of the Niger and Orinoco deltas (Porrenga 1965) is consistent with this interpretation.

Another widespread belief is that the ironstones did not form in the open sea but in some sort of near-shore lagoon. To be meaningful, the term lagoon should be restricted to partly isolated regions of abnormal salinity, or at least regions protected from strong wave- or tide-induced water disturbance. The Jurassic ironstones contain abundant benthonic fossils

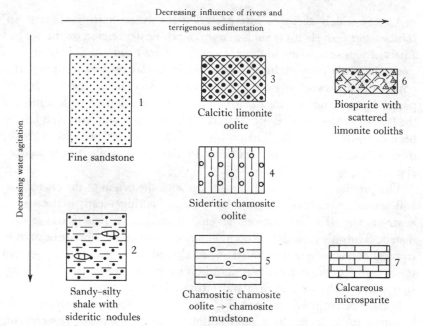

FIGURE 3.5 Proposed environmental conditions controlling the formation of the Upper Pliensbachian ironstones and associated deposits in Great Britain (Hallam 1967 c).

including many stenohaline types, with no suggestion of abnormal salinity. There is, in addition, abundant evidence from the rocks themselves of strong water agitation. The notion that the ironstones possess a fauna distinctive from other shallow water deposits has not been borne out by recent work (e.g. Hallam 1963 a, 1967 c; Brookfield 1973). The Westbury Ironstone of the Upper Oxfordian may be an exception, and Talbot (1974) has argued a reasonable case for a lagoonal origin on the basis of its restricted fauna, largely oysters, and evidence of quiet water deposition.

Different solutions have been proposed to account for the general rarity of detrital quartz and feldspar grains and normal illitic clay. Hemingway (1951), influenced by an approximation in the Yorkshire Lias on three occasions to a cyclic sequence shale–sandstone–ironstone (see fig. 3.4), has related the sedimentation to erosional cycles on the land. As the latter became peneplaned, mechanical erosion diminished so that finally only chemical derivatives, including iron, were transported and deposited in shallow restricted arms of the sea. This hypothesis fails to account for the frequent lateral passage of ironstones into clastic rocks and is clearly incompatible with the interpretation adopted here.

A more widely accepted interpretation involves a sort of 'clastic trap' (Huber and Garrels 1953) such as a topographic depression on the sea bed adjacent to a submarine high or 'swell' where a chemical deposit such as ironstone could form undiluted by terrigenous clay or sand. Although this popular hypothesis has been challenged by Brookfield (1971) it is supported by the tendency for the ironstones to be condensed with respect to the laterally correlative clastic deposits. This is consistent with their having been deposited on swells. The problem remains, of course, how to get the iron-bearing sediments or solutions to the swell while the associated clastics are, as it were, filtered off.

The problem is very much bound up with the origin of the chamosite, which it is generally agreed must have formed authigenically in the marine environment. It occurs as well-developed ooliths, as pellets and as mud matrix. The only records of Recent chamosite are in the submarine parts of tropical deltas, where its occurrence as altered faecal pellets suggests that the presence of organic matter might be a necessary prerequisite (Porrenga 1965), and to the west of Scotland (Rohrlich *et al.* 1969).

It has been widely accepted that iron was transported in rivers in the ferrous state protected by organic colloids, and later deposited on entering the alkaline environment of the sea. Carroll (1958) has put forward a different suggestion which seems, on the face of it, very plausible. She pointed out that ferric oxide can travel to the sea as particles coating the surface of clay micelles and can be removed by a lowering of the redox potential to negative values, which converts the ferric iron to the more soluble ferrous form. It is argued that chamosite might be formed in such a reducing environment by the interaction of iron, released in this way, with kaolinite (cf. Schellmann 1969).

It seems quite probable that siderite could form as a result of the process Carroll described, iron released from clay minerals within silty prodelta muds reacting with carbonate ions in the interstitial water and afterwards segregating to form nodules. Indeed, diagenetic siderite is forming today in muds of the Mississippi delta region (Ho and Coleman 1969). Unfortunately, Carroll's interpretation does not solve the problem of chamosite concentrations. All the normal marine Jurassic clays consist, as already noted, dominantly of illite, with kaolinite being markedly subordinate. Yet chamosite is the most abundant clay mineral in the ironstones. The influx of large quantities of illite, just as with detrital quartz, evidently diluted the available iron to such an extent that exploitable ore could not form. Therefore, in a sense we are back where we started. It is hard to resist the conclusion that some iron was carried, perhaps as a colloidal solution or suspension, beyond the area where most of the terrigenous sediment was deposited.

Chamosite presumably formed authigenically in a reducing environment within fine-grained sediment (cf. Curtis and Spears 1968) but was evidently stable in slightly oxidising conditions, as testified by the rich benthonic faunas that are often present. Presumably it was carried by burrowing organisms to the surface, where ooliths formed. It is possible that planktonic organisms such as diatoms could have contributed to both the iron and silica required for the formation of chamosite, since opaline silica is readily soluble within sediments. As regards the difficulty concerning the lack of close modern analogues, one should bear in mind the following point. Because the world today probably has abnormally high relief, terrigenous sedimentation rates on the continental shelf in the vicinity of river deltas are probably a good deal higher than in the shelf sea that covered northwest Europe in the Lower and Middle Jurassic. The concentration of iron coming from rivers therefore need not have been appreciably higher than in comparable climatic regimes today, so long as sedimentation rates were slight.

The formation of the different types of ooliths poses other intriguing problems. Given the initial production of chamosite ooliths, several types could develop. In relatively low energy environments, if sufficient iron were available in the interstitial water to allow the formation of siderite, the ooliths might have remained in their original spherical or subspherical condition. If not, then they would have suffered compaction and colloidal shrinkage of their interiors to produce irregular *spastoliths*, which often exhibit hooked junctions of their outer rinds indicating their original size and shape. In high energy environments, the ooliths might have been partially oxidised to goethite, though some clay mineral would remain, or be broken in two, with the fragments sometimes serving as nuclei for a later generation of ooliths (Hallam 1967 c).

The original ooliths are generally similar in shape and size to aragonite ooliths and probably formed in a quite similar way, rather than having grown colloidally within the sediment as envisaged by Caillère and Kraut (1954) for the Lorraine ores. The analogy with chamosite pisoliths in bauxite is misleading because these have grown to much greater sizes. Certain special features should be stressed, however. Oolitic beds with the highest iron content are often best developed at the centres of sedimentary basins rather than at the margins, and the ooliths might attain their greatest size in beds with a matrix of chamositic mud rather than sparry calcite, suggesting an inverse relationship with degree of agitation and winnowing activity of the bottom waters. Nuclei are not always conspicuous: often they may have consisted merely of small flakes of chamosite.

An illuminating and admirably detailed study of chamosite oolith

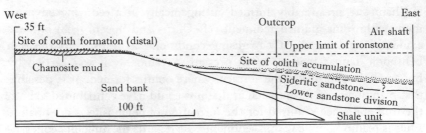

FIGURE 3.6 Proposed relationship of the Winter Gill ironstone (Middle Jurassic, Yorkshire) to the site of oolith accumulation (Knox 1970).

formation in a thin Middle Jurassic ironstone in Yorkshire led Knox (1970) to the conclusion that the ironstone bed was the site of accumulation but not of formation of the ooliths. These had been transported by bottom currents from a nearby shoal, where they had formed on a surface of chamosite mud under predominantly low energy conditions (fig. 3.6). The fact that the ironstone lies only a metre below a rootlet horizon suggests deposition in extremely shallow water.

PALAEOECOLOGICAL ASPECTS

The invertebrate fauna also provides evidence bearing on the nature of the environment, most usefully, perhaps, in relation to depth and salinity of the sea. With regard to the former, the faunal evidence may be said to reinforce that from the sediments, but in the case of salinity the fossil data are the more significant.

The shallower water deposits, characterised by sandstones, sandy shales and oolitic ironstones, tend to be relatively rich in bivalves (usually the dominant invertebrates), often quite large in size, which in life were suspension feeders. To give an example from the Lias, this holds both for surface dwellers such as *Gryphaea, Pseudopecten, Plicatula* and *Modiolus* and for deep burrowers such as *Pholadomya* and *Pleuromya*. Bivalves in the deeper water facies, characterised by smooth-textured shales and mud-stones, tend to occur more sparsely and are generally small in size, sometimes with a high proportion of junveniles. Deposit-feeding proto-branchs such as *Nuculana, Mesosacella, Palaeonucula* and *Rollieria* are more common in this facies, which also has small forms of such genera as *Astarte, Bositra* and *Luciniola*.

Ammonites, on the contrary, are commoner and more diverse in the deeper water facies. Their comparative rarity in the shallower water facies is not attributable to preservation failure, because formerly aragonitic-shelled bivalves and gastropods are well preserved as moulds, as are those ammonites which do occur.

Among the other shelly fossils, crinoids deserve a mention, as being more abundant in the shallower water facies, where the disarticulated ossicles frequently constitute a major component of the more condensed beds such as the ironstones. Brachiopods are likewise commoner in this facies.

Since the change from one facies to the other often takes place over very short distances geographically at the same horizon, or in vertical sequence, as in the types of small-scale sedimentary cycle described by Sellwood (1970), the differences in depth of sea must have been slight, of the order of a few metres or tens of metres. The faunal differences among the bivalves are most plausibly accounted for by differences of substrate, the deeper water muds being more soupy and unsuitable for suspension feeders but providing plenty of food for deposit feeders (cf. Rhoads and Young 1970), while temperature and salinity fluctuations probably discouraged the ammonites from entering in large quantities the shallowest waters of a few metres depth (Hallam 1972*a*).

Trace fossils belonging to several ichnogenera occur commonly in the Jurassic of northern Europe and attempts have been made to use them as bathymetric indicators. Thus Farrow (1966) thought he could recognise in certain Middle Jurassic marine beds on the Yorkshire coast (the Scarborough Beds) a series of trace fossil depth zones, using independent sedimentary and faunal criteria as indicative of southward deepening of the sea bed. From shallowest to deepest, Farrow distinguished successively: 1, parallel-oriented *Thalassinoides*; 2, parallel-oriented *Rhizocorallium*; 3, unoriented *Thalassinoides*; 4, unoriented *Rhizocorallium*; 5, *Thalassinoides* and *Asterosoma*. The oriented trace fossils are supposed to signify tidal action and probably the intertidal zone.

Although well illustrated and stimulating, Farrow's paper is open to criticism on several grounds. Independent examination of the beds in question raises doubts about the validity of the distinction between the supposedly parallel-oriented and unoriented *Thalassinoides* and *Rhizocorallium*. Bearing in mind Seilacher's (1967) familiar depth zonation one wonders why the shallowest water, purportedly intertidal, beds do not contain vertical burrows such as *Diplocraterion*.

A more plausible bathymetric subdivision is made by Ager and Wallace (1970) for the Kimmeridgian sandstones and shales of the Boulonnais in northwest France. In their scheme four depth zones were distinguished (fig. 3.7):

1. Deeper water, flat-bedded muddy sediments – horizontal *Rhizocorallium*.

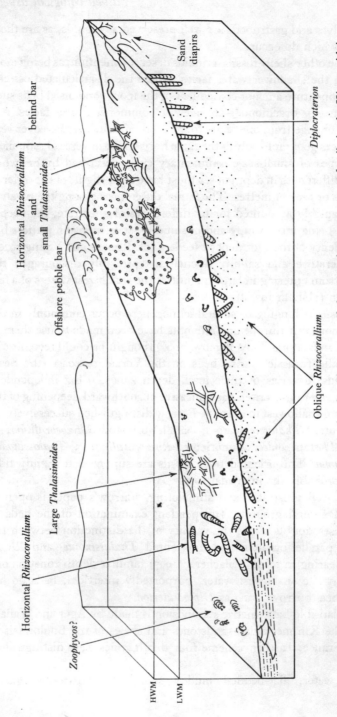

FIGURE 3.7 Block diagram showing the postulated environments of the more important trace fossils in the Kimmeridgian of the Boulonnais (Ager and Wallace 1970).

Sand diapirs

Diplocraterion

Horizontal *Rhizocorallium* and small *Thalassinoides* } behind bar

Offshore pebble bar

Oblique *Rhizocorallium*

Large *Thalassinoides*

Horizontal *Rhizocorallium*

Zoophycos?

HWM

LWM

2. Shallow water, high energy, thick-bedded sediments – large *Thalassinoides* (locally *Ophiomorpha*).
3. Subtidal, thin-bedded, ripple marked sandy sediments – *Rhizocorallium* oblique to bedding.
4. Intertidal high energy sandy sediments with erosion levels – vertical *Diplocraterion* and other deep burrows.

Locally, in low energy intertidal or shallow subtidal sediments small *Thalassinoides* occur. *Thalassinoides* characterises calcareous and argillaceous sediments, *Ophiomorpha* sandy sediments; this suggests that the differences are preservational and that the same burrowing crustacean was responsible. In contrast to Farrow, no preferred orientations were detected. Zones 1 and 2 are compared with Seilacher's (1967) *Cruziana* and 3 and 4 with his *Glossifungites* and *Skolithos* Facies. Ager and Wallace acknowledged, however, that the trace fossil distribution relates to water energy level rather than bathymetry *per se*.

A further attempt at bathymetric zonation has been made by Wincierz (1973) for a sequence of sandstones and sandy shales in the Hettangian of northwest Germany. The shallowest facies, held to correspond to Seilacher's *Skolithos* Facies, is characterised by *Diplocraterion*, and the deepest (but still shallow water) facies by horizontal *Rhizocorallium* – a scheme which corresponds quite well with that of Ager and Wallace. *Thalassinoides* occurs commonly throughout the sequence and thus has no depth significance within the shallow water regime.

More generally for the Jurassic, one can state that the shallower water marine sediments are characterised by abundant *Thalassinoides* and *Rhizocorallium*, and the shallowest, perhaps locally intertidal, deposits by U-burrowers such as *Diplocraterion* or *Arenicolites*. *Thalassinoides* and *Rhizocorallium* tend to occur in different beds, but it is far from evident that this is a simple matter of depth difference. The tendency towards mutual exclusion may reflect a subtler environmental variable, or some form of ecological competition between the crustaceans responsible for these burrowing systems. In the deeper water shaly deposits these two trace fossils are much rarer, and the only common types are *Chondrites* (which also occurs in shallower water) and pyritised sinuous, irregularly oriented cylindrical structures attributable to burrowing worms.

With regard to the faunal evidence for salinity changes, a very full palaeoecological analysis involving body fossils was undertaken by Hudson (1963) on the Great Estuarine Series (Bathonian) of the Scottish Hebrides. The evidence of the sediments suggests an environment of muddy lagoons into which small river deltas extended. A variety of features indicated a brackish water environment. Thus the total number of species of macro-invertebrates is less than fifty, which is far fewer than for the marine stratigraphic equivalents in southern England. Furthermore, a few metres of one unit, the Lower Ostrea Beds, accounts for nearly all the records of stenohaline organisms such as echinoids, bryozoans and brachiopods; corals and cephalopods are entirely absent. The fauna which is present includes some well-known brackish water forms. Finally, the sediments include drifted plant beds but no evaporites; marine beds occur above and below the beds in question.

Hudson drew a comparison with the coastal bays of Texas at the present day, which exhibit strong salinity fluctuations from mainly brackish to occasionally hypersaline over short time periods. Lunachelles of the Bathonian *Liostrea hebridica* and *Praemytilus strathairdensis* were compared with modern reefs of *Crassostrea virginica* and *Brachidontes*. A salinity series of fossils in the Great Estuarine Series was erected (fig. 3.8):

1. Least saline: *Unio, Viviparus, Neomiodon, Euestheria*.
2. Intermediate: *Praemytilus* and *Liostrea* dominant.
3. Most saline (slightly brackish): *Liostrea, Modiolus*, rhynchonellids etc.

In a briefer palaeoecological account, as part of their description of the Lower Lias of the Lossiemouth borehole in northeastern Scotland, Berridge and Ivimey-Cook (1967) also noted faunal evidence for gradations in salinity in sediments of typical 'Coal Measure' type. The marine strata have a varied, moderately diverse fauna of bivalves with a few types of ammonite. A slightly brackish environment is signified for other strata by an absence of ammonites but the presence of *Lingula* together with several bivalve genera, such as *Astarte, Gervillella, Liostrea, Modiolus* and *Protocardia*. The least saline deposits are characterised by *Euestheria* and darwinulid ostracods.

Jurassic Foraminifera may also prove useful as salinity indicators. According to Kaptarenko-Tschernousowa (1964), marginal marine clastic sediments of the type that pass into coal measures are characterised by a few genera of agglutinating foraminiferans of simple structure, such as *Ammobaculites, Ammodiscus, Hyperammina* and *Rheophax*.

Jurassic trace fossils have been much less discussed from the point of view of salinity, but the presence or absence of certain types provides useful information. The absence of ichnogenera such as *Chondrites, Diplocraterion, Rhizocorallium* and *Thalassinoides*, so abundant in shallow water beds containing body fossils of stenohaline groups, may be indicative of abnormal salinity, and indeed they appear to be absent from or very rare in presumed non-marine rocks in the Middle Jurassic of Yorkshire and

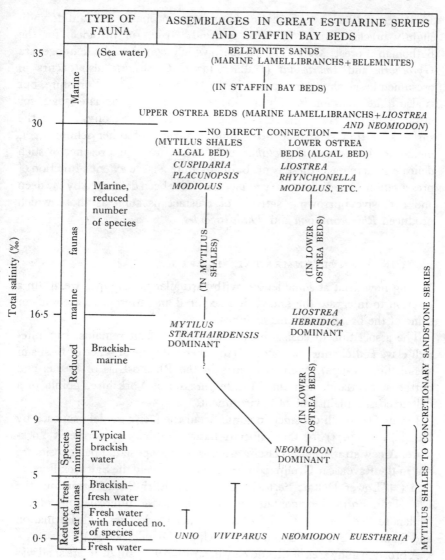

FIGURE 3.8 Salinity-controlled fossil assemblages in the Bathonian and early Callovian of the Skye area, Scotland (Hudson 1963).

Scotland, and the topmost Jurassic–basal Cretaceous 'Purbeck Beds' in Dorset. It is harder to characterise brackish water deposits by the presence of distinctive ichnogenera, but the irregular burrow fillings known as *Planolites* are common in parts of the rock units cited above.

Certain deposits consisting of small-scale alternations of shale and sandstone and lacking stenohaline faunas were probably also laid down in slightly brackish water. This seems to be true, for instance, for the Bathonian Forest Marble of Dorset, which contains the ichnogenera *Gyrochorte* and *Imbrichnus* (Hallam 1970). *Gyrochorte* also occurs in presumed brackish water deposits in the Middle Jurassic of Yorkshire, but is also quite common in fully marine rocks of the appropriate facies, for instance in the Hettangian of Swabia and Lower Saxony, where the surfaces of thin sandstone beds show a variety of other ichnogenera, including *Asteriacites*, *Curvolithus* and *Neonereites*. The presence of such delicate structures as these is probably to quite a large extent a function of preservational conditions; they must easily have been destroyed by the deep and extensive burrowing activity of crustaceans such as those which produced *Rhizocorallium* and *Thalassinoides*.

REGIONAL ENVIRONMENTAL MODELS

Having now dealt at some length with particular facies types we are in a position to integrate the knowledge acquired and attempt to account for some of the broader regional changes.

The association of coals, shales with drifted plant remains, kaolinite-rich clays and channel sandstones (together with the absence of fossils of stenohaline groups), such as occurs in the Rhaeto-Lias of Scania and northeast Scotland, and the Middle Jurassic of Yorkshire, points to a deltaic coastal plain type of environment.

Let us first of all consider in some detail the facies model proposed by Hemingway (1974) for the Bajocian–Bathonian delta of northeast Yorkshire. A new stratigraphic classification is now proposed. The deposits are put in the Ravenscar Group, comprising successively the Saltwick Formation (≡ 'Lower Deltaic Series'), the marine Eller Beck Formation, the Cloughton Formation, containing the thin marine intercalations of the Millepore Bed and Whitwell Oolite, the marine Scarborough Formation and finally the Scalby Formation (≡ 'Upper Deltaic Series').

A broad marshy area of silt and clay deposition with subordinate sand was interrupted locally by fresh water lagoons with unionid bivalves; thickets of the horsetail *Equisetites* sometimes formed thin peat accumulations. Elsewhere ferns and cycadophytes colonised the dry areas. The marsh surface was overwhelmed periodically by coarser-grained fluvial sediments, during relatively emergent phases. Major distributaries overspread the area, depositing medium- to coarse-grained cross-bedded sandstones and cutting channels up to 30 m deep. Thus channels in the basal Saltwick Formation locally reach down through the marine Dogger sandstones into

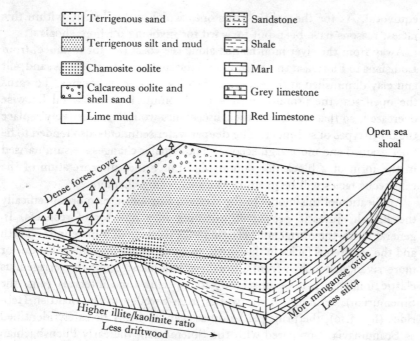

Terrigenous sand		Sandstone	
Terrigenous silt and mud		Shale	
Chamosite oolite		Marl	
Calcareous oolite and shell sand		Grey limestone	
Lime mud		Red limestone	

FIGURE 3.9 Simplified model of the proposed palaeogeographic relationships between the major facies of the European Jurassic. (Adapted from Hallam 1967b.)

the Upper Lias clays. The Moor Grit of the Scalby Formation is a single fluviatile sheet. Much of the sand is probably derived from terrestrially exposed Carboniferous sediments to the north. During submergent phases there were incursions of shallow sea, predominantly from the south, with the deposition of shales, sandstones and limestones with a marine fauna of restricted diversity, dominated by bivalves.

Returning to a consideration of more general facies relationships, the widespread association of thin, often bioturbated, fine sandstones and sandy shales with moderate kaolinite content and layers of siderite nodules, and containing a moderately diverse marine fauna, is less easy to pinpoint precisely. Such beds are clearly associated with the deltaic group and were evidently deposited at no great distance offshore from river mouths, and partly at least in that submerged portion of the sedimentary accumulation known as the prodelta. Offshore bars, intertidal flat and beach deposits may occur locally but are not generally easy to identify with confidence, both because the limitations of outcrop and borehole data usually preclude reliable detailed three dimensional reconstructions of the facies, and because interpretation of the sedimentary structures may be

53

equivocal. As for the oolitic ironstones which occur locally within this facies, reasons have been put forward for invoking offshore shoals.

Away from the river mouths but along the coast (as, for instance, from Louisiana to Florida at the present day) the content of terrigenous sand, silt and clay diminished at the expense of shallow water carbonates. Towards the open sea, the proportion of sand, kaolinite and driftwood likewise decreased, so that marls and then limestones would progressively replace the other types of sediment. The deeper water sediments also tended to be finer grained and more argillaceous. These major changes are summarised in the form of a block diagram in fig. 3.9. Fuller interpretation of the calcareous facies will be given in the succeeding chapters.

Application of this model to the western European, or more specifically the British, Jurassic proves illuminating in a number of respects. In general terms the terrigenous clastic facies tends to dominate in the north and the calcareous facies in the south, from which we may infer that one or more rivers debouched from a northern landmass. Since the terrigenous clastic influence can be observed, for specific time intervals (e.g. the Sinemurian), to diminish westwards from Sweden to the Scottish Hebrides (fig. 3.10), the principal landmass can probably be safely identified as Scandinavia. Compared with the Sinemurian, the early Pliensbachian exhibits a broadly similar facies pattern, but the offshore facies tends to be more calcareous and the coal measures in the north were replaced by the inshore marine clastic facies. These changes can simply be accounted for by a modest rise of sea level (e.g. Sellwood 1972b). The late Pliensbachian, in contrast shows a facies distribution indicative of regression compared with the deposits above and below. At least one new source of clastic sediment must be invoked to account for the influx of sands in southwest England, and presumably relates to tectonic uplift of land. By contrast, the Lower and Middle Toarcian mark a renewed transgressive phase, with deeper water shales replacing shallower water sandstones and ironstones, and ferruginous sediments diminishing at the expense of calcareous (Hallam 1967c); see figs. 3.11 and 3.12.

Exploration for hydrocarbons in the North Sea has revealed the existence of a structure, extending eastward from southern and eastern Scotland and known as the Mid North Sea High, where lower Tertiary rocks are directly underlain by pre-Mesozoic basement. This structure was probably produced as a result of intra- or post-Cretaceous uplift and hence cannot be treated with much confidence as a positive area in the Jurassic, like the London–Ardennes massif. Most or all of the Jurassic sand was probably reworked from late Palaeozoic sandstones in the North Sea region but there is no evidence that the Mid North Sea High was in existence in

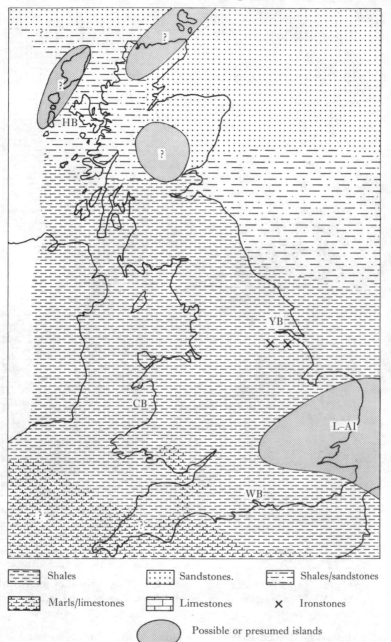

FIGURE 3.10 Tentative facies map for the Sinemurian of the British area. CB = Cardigan Basin, WB = Wessex Basin, YB = Yorkshire Basin, HB = Hebridean Basin, LAI = London–Ardennes Island.

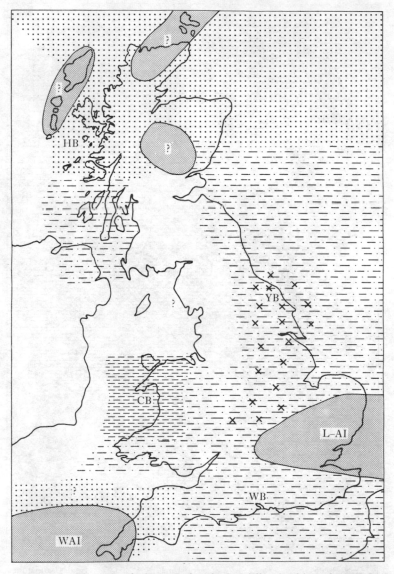

FIGURE 3.11 Tentative facies map for the late Pliensbachian of the British area. W–AI = Western Approaches Island. Key as in fig. 3.10.

Liassic times and hence it is not included among the postulated land-masses portrayed in figs. 3.10 to 3.12. The palaeogeographies portrayed in these figures, and in fig. 3.13, must be regarded as extremely speculative because there is very little critical data and because of the considerable

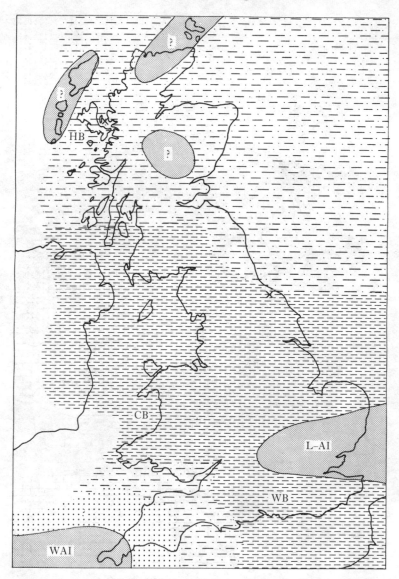

FIGURE 3.12   Tentative facies map for the Toarcian of the British area.
Key as in fig. 3.10.

amount of pre-Tertiary regional uplift in northern Britain and the North
Sea region. The illustrations are meant essentially to portray the distribution of sedimentary facies.

At the commencement of the Middle Jurassic regional tectonic uplift in

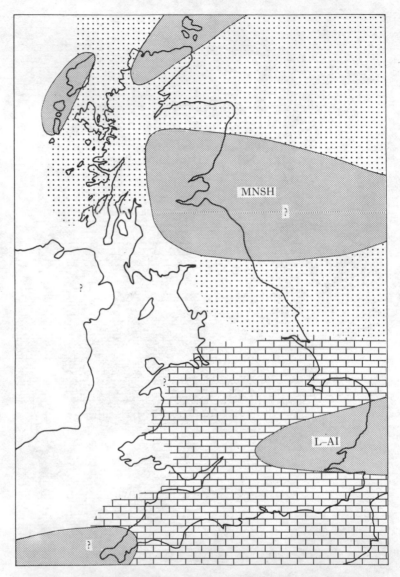

FIGURE 3.13   Tentative facies map for the Bajocian of the British area.
MNSH = Mid North Sea High. Key as in fig. 3.10.

England had the effect of causing the southward spread of deltaic facies into Yorkshire, where it persisted with only minor marine floodings through the Bajocian and Bathonian. Facies changes in the Bajocian Eller Beck and Scarborough formations (Knox 1973, Farrow 1966) signify a landmass not far to the north or northwest, which might well mean that the Mid North Sea High and its landward continuation began to emerge at this time. Contemporary with these changes, the predominantly muddy region of southwest England became converted into an extremely shallow platform on which oolites and other types of limestone were deposited. The East Midlands mark the area of interdigitation of the two facies, both deposited very close to sea level.

The Callovian marks the widespread transgression of more offshore and deeper water facies in almost every region, which clearly reflects a relative rise of sea level. A widespread argillaceous facies persisted until the late Oxfordian, when the calcareous facies of southern Europe extended northwards as far as northern England and Poland. This change need not simply have resulted from a regional shallowing: it could also signify a marine transgression, with the more offshore calcareous facies spreading at the expense of the more clastic near-shore, or more correctly near-delta facies. Careful and thoughtful work is required in particular areas to disentangle these different factors.

The examples cited above should be sufficient to illustrate the utility of the model. It signally fails to explain, however, the distribution in space and time of the distinctive facies of laminated bituminous shales. Why should such beds have spread at times over large parts of western Europe?

One important clue is that the widespread horizons characteristically occur close to the base of marine transgressions. This is true, for instance, of the early Hettangian and Toarcian and the middle Callovian deposits, as also of the Rhaetian black shales and indeed many non-Jurassic examples of similar lithology. This fact, together with their widespread occurrence in the midst of undoubted shallow water deposits, rules out a straightforward comparison with a deep barred basin like the Black Sea (fig. 3.14(*a*)). Such a comparison has frequently been made in the case of the German Posidonienschiefer, though depths of the order of hundreds rather than thousands of metres are invoked (e.g. Brockamp 1944). Recently the discovery of coccoliths in both the Posidonienschiefer and the Black Sea bituminous muds has led Müller and Blaschke (1969) to support the comparison, but of course the mere presence of coccoliths tells us something about pelagic influence but nothing about depth. Furthermore, the pelagic influence is not necessarily paramount, because palynological investigation of the correlative Jet Rock in Yorkshire suggests the likelihood of an inshore environment (Wall 1965).

It should be borne in mind that conditions at the present time may in certain respects be an unreliable guide to the past. For example, all the Recent oceans are well ventilated at depth as a direct consequence of the climate, with a strong circulation of oceanic currents depending on a pronounced temperature gradient between the tropics and poles. In the more equable periods of the past, such as the Jurassic (see chapter 9), much of the ocean bottom could have been more or less stagnant. A tendency towards stagnation is far more widespread in shallow tropical and subtropical waters than in their temperate and polar equivalents because of the higher organic productivity and rates of oxidation in the former; such conditions could have been almost worldwide in the Jurassic. Brongersma-Sanders (1971) has indeed argued that restricted circulation as a control on bottom stagnation has been greatly overemphasised at the expense of organic productivity. We can also note that the strength of winds, as of ocean currents, is ultimately dependent on latitudinal temperature gradients. Therefore during equable periods winds might have been weaker than today, with the depth of effective wave disturbance being correspondingly reduced.

The character of the litho- and biofacies in western Europe point clearly to the existence of a shallow epicontinental sea flanked by land of low relief. Though the bituminous shales cannot seriously be regarded as deep water deposits, nor can they be compared with lagoonal muds where circulation is restricted by 'meadows' of benthonic algae. Not only would fine lamination fail to develop in such deposits, they should contain a rich fauna of shelly epibionts which formerly lived on the algal fronds above the anaerobic muds (cf. Bauer 1929). There is no suggestion of this from the shale fauna; nor is there any indication of abnormal salinity.

Keulegan and Krumbein (1949) have shown mathematically that, given a sufficiently low oceanward gradient of the sea bed, a condition will be attained whereby waves generated by the wind some distance offshore will dissipate their energy before reaching the shore. Not only would the formation of beaches and cliffs be prevented but extensive stretches of sediment on the sea bed would be subjected to very little wave disturbance. Furthermore, tidal range and currents should also be damped significantly. Whereas present shorelines have a seaward slope usually in excess of 1 in 100, the model utilised by Keulegan and Krumbein had a beach slope of 1 in 600, diminishing considerably offshore.

One can therefore propose an interpretative model for the Jurassic bituminous shales whereby some restriction of water circulation was a consequence of the geometry of the epicontinental sea, an extensive sheet of shallow water being very little agitated below wave base (fig. 3.14(*b*)). An abundance of vegetable matter, in what was in effect a fairly inshore area subjected to a warm, humid climate, was rapidly oxidised, so favouring total or partial stagnation whenever water circulation was minimal. As the

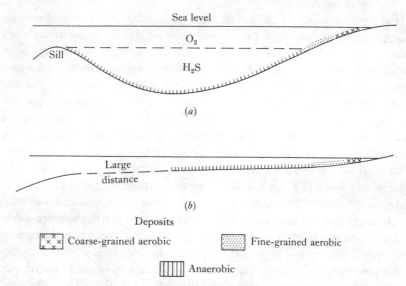

Sea level

O_2

Sill

H_2S

(a)

Large
distance

(b)

Deposits

Coarse-grained aerobic   Fine-grained aerobic

Anaerobic

FIGURE 3.14   Alternative environmental models for Jurassic bituminous shales:
(a) barred basin, (b) shallow shelf sea (Hallam 1967c).

sea deepened during the course of the transgression so freer connections
were established with the circulation of the open sea, so that eventually
bottom aeration was sufficient to support a moderate benthonic life. Thus,
for example, the Middle Toarcian Alum Shales of Yorkshire are, on this
interpretation, deeper water deposits than the underlying Jet Rock.
Applying the Black Sea model to the latter, Hemingway (1951) naturally
concluded that the Alum Shales were laid down in shallower water.
Likewise, the more marly Upper Oxford Clay of southern England might
have been laid down in deeper water than the bituminous Lower Oxford
Clay.

The interpretation that the laminated bituminous shales are shallow
water deposits is also favoured by the existence of numerous small-scale
alternations of these and shelly, non-laminated marls and limestones that
occur, for instance, in the Blue Lias (Hallam 1960). One can only
speculate about the likely depth of deposition. There appears to be little
significant wave abrasion in the sea today below about 10 to 15 m. Bearing
in mind the probably very modest oceanward gradient of the bottom and the
likelihood of relatively weak winds, a figure of 15 to 30 m might be
considered a reasonable guess. This allows for the possibility of periodic
storms disturbing water to a greater depth than usual, while recognising
that an appreciably greater depth would render likely a freer connection
with the Tethyan Ocean.

It cannot be assumed, however, that all non-bituminous, bioturbated shales with benthonic fauna are necessarily of deeper water facies than associated bituminous shales. Indeed, one shortcoming of such an interpretation is that deposits such as the Middle Toarcian Alum Shales and Lower Oxfordian Oxford Clay pass up gradually into obviously shallow water sands without the intervention of another bituminous shale sequence. Furthermore, the Kimmeridge Clay of Dorset is a thicker, more basinal and deeper water seqence than the sandier equivalent in the Boulonnais, yet it has a much higher proportion of laminated bituminous shale intercalated as thin units alternating with more normal shale. In a similar way the more basinal facies of the Lower Pliensbachian in the eastern Cantabrian Mountains of northern Spain is more bituminous than the thinner sequence further south (Dahm 1966). Nevertheless there seems no good reason to doubt that the bituminous shales of northwest Europe are relatively shallow water deposits, for which the deep 'barred basin' model is inapposite, and the occurrence of many such deposits close to the base of transgressive sequences is more readily explicable along the lines portrayed in fig. 3.14(b). Clearly each particular case must be treated on its merits, with a full consideration of the lateral and vertical stratigraphic relationships.

The notion that extremely low oceanward gradients, such as probably characterised the European Jurassic epicontinental sea, might act to damp both waves and tides, with important consequences for the sedimentary environment, has been taken up enthusiastically by both Shaw (1964) and Irwin (1965). If we consider modern examples, the frictional effect of the wide shelf of the Bering Sea appears to exert such a damping effect, in association with surface ice, that tidal range at the coast is reduced almost to zero, and diurnal tides over the Great Bahama Bank and in Florida Bay are also negligible. On the other hand, another wide shallow epicontinental sea, the Yellow Sea, has an appreciable tidal range. Much evidently depends upon the shape of the marine basin and upon tidal resonance factors. Generalisations applicable to epicontinental seas of the past are therefore difficult to make. Returning to the European Jurassic, the model adopted to account for the widespread bituminous shale horizons would appear to favour a Shaw–Irwin type of interpretation involving tidal damping, but Sellwood (1972a) records sedimentary data which are at least strongly suggestive of tidal action. Unfortunately such deposits are not widespread in space or time, nor is there any reliable means of assessing tidal range. Further points of difficulty in recognising tidal deposits of the past are, firstly, that many features such as clay drapes, flaser bedding and herringbone cross lamination may not be confined to regimes with an appreciable tidal range even at the present day (the 'experimental control' of establishing their absence from regions such as the Rhône delta has not yet been carried out). Secondly, there seems no means of distinguishing in the stratigraphic record between the effects of diurnal tides and those more infrequent and irregular wind-generated 'tides' known as seiches.

4 Calcareous and associated argillaceous facies in northern and central Europe

Limestones and marls occur widely to the north of the circum-Mediterranean province, extending far into England in the Bajocian–Bathonian and late Oxfordian, but until recently very little was known in detail about their environments of deposition. Within the last ten years or so considerable advances have been made in our understanding of these rocks, which relates in large measure to our greatly improved knowledge of Recent carbonate deposits in the Bahamas and the Persian Gulf. It is desirable to discuss certain types of deposit and aspects of the palaeo-ecology irrespective of location and stratigraphic position before going on to consider several examples of comprehensive facies analysis. Our present state of knowledge precludes the type of regional stratigraphic survey attempted at the end of chapters 3 and 5.

LIMESTONE TYPES
Algal limestones
Although the remains of calcareous algae were spotted in the Lower Purbeck Beds of Dorset many years ago the first adequate description dates from only recently (Pugh 1968). Pugh recognised traces of blue–green, green and red algae. The blue–greens include both discrete colonies occurring as nodules (*Girvanella*) and former algal mats, with calcareous laminae produced by sediment entrapment. The latter fall into three categories of stromatolites/oncolites, as distinguished by Logan, Rezak and Ginsburg (1964).

The first category, 'laterally linked hemispheres' (LLH), comprises the tufaceous limestones of the basal Purbeck unit known as Hard Cap, with concentric structures developed around cylindrical hollows thought to represent the site of decayed wood. Secondly, 'stacked hemispheres (SH) are recognisable in the well-known 'burrs' of Soft Cap, best seen at the so-called Fossil Forest to the east of Lulworth Cove. These structures, ranging up to 2 m in diameter, likewise envelop central depressions left behind after the decay of 'boles' or tree stumps. Thirdly, oncolites occur in the Cypris Freestones.

The green algae belong to the Codiacea and include the genera *Ortonella* and *Cayeuxia*, which occur as discrete nodules; the red algae are represented by *Solenopora*.

A detailed investigation by Hudson (1970) of several algal limestone horizons in the Great Estuarine Series of the Scottish Hebrides proved very rewarding in terms of environmental interpretation. The Lower Ostrea Beds contain horizons with algal nodules belonging to *Cayeuxia* and other genera. These are thought to signify supratidal conditions because early calcification and cementation is necessary to ensure the preservation of the algal tubules, which are normally destroyed within subtidal stromatolites.

The stromatolitic, algal mat deposits contain pseudomorphs of calcite after gypsum. Together with the presence of desiccation pores in the limestones, this strongly suggests periodic exposure of algal mats, with evaporation of water trapped among the algal heads. A comparison was made with deposits in the Laguna Madre of Texas, where gypsum crystals have been recognised within algal mats. This hypersaline lagoon lies to the southwest of those brackish coastal bays which Hudson (1963) had earlier invoked as an environmental model for the Great Estuarine deposits as a whole.

An algal origin has very plausibly also been invoked for those carbonate-rich limonitic concretions known colloquially in Dorset as 'snuff-boxes'. These structures have concentric laminae, are generally discoidal in shape, and range in size up to 30×10 cm; they are commonly encrusted by serpulids and association with stromatolitic layers and bored and limonite-stained pebbles. Occurrences in the Aalenian and Bajocian of Dorset and Somerset and the Bajocian of Normandy were carefully described by Gatrall *et al.* (1972) (see also Fürsich (1971) for the Normandy occurrences). Moderately strong water movements are invoked to account for the periodic rolling over of the concretions to ensure an oncolitic pattern of growth. The limonite concentration is thought to be partly algal and partly inorganic in origin.

Such oncolites and the accompanying stromatolites are clearly associated with highly condensed sedimentary sequences and hardgrounds, as established also by Szulczewski (1968) in his comprehensive description of Middle and Upper Jurassic stromatolites in the Holy Cross Mountains of southern Poland. Other occurrences of limonitic algal concretions are in the Pliensbachian of Normandy and the Callovian of the Swiss Jura Mountains.

There are no reasonable grounds for invoking subaerial emergence in the intertidal zone to account for the stromatolitic laminae that occur in the highly condensed Lower Toarcian part of the so-called Junction Bed in west

65

Dorset (Sellwood *et al.* 1971). This pale pink, yellow-weathering micritic limestone contains abundant ammonites but few benthonic fossils, and was without much doubt laid down in deeper water than the other types of algal limestone discussed. (In fact it compares quite closely with a certain type of deposit much more characteristic of the circum-Mediterranean belt than northern Europe.) Hence the stromatolites are likely to be of subtidal origin, like those off Andros Island and Bermuda, described respectively by Monty (1965) and Gebelein (1969).

Hardgrounds

Within the more condensed limestone sequences there appear at quite frequent intervals beds or bedding surfaces that give indications of cementation penecontemporaneous with sedimentation. The evidence that proves contemporary hardening comes from encrusting organisms, most commonly oysters and serpulids, rock borers including bivalves of the genus *Lithophaga*, which excavate sac-like hollows, and the narrow-bore tubes of polychaete worms. Hölder and Hollmann (1969) have examined the polychaete borings in the Bajocian of the English Cotswolds and the Bathonian of the Boulonnais, and distinguished two types, the U-shaped *Polydorites* and the more or less straight vertical or oblique *Potamilla*-type. In contrast to burrows excavated in soft limy sediment, the sediment fillings of the borings, normally brown marl in weathered outcrops, exhibit a sharp physical and chemical contrast with the matrix. (One can also sometimes find crustacean excavations like *Thalassinoides* in hardgrounds, filled with marl, which indicate early cementation of the burrowed sediment.)

Serpulids are occasionally found encrusting *undersurfaces* of hardgrounds, proving excavation of soft sediment from beneath a surface layer of cemented material (Purser 1969). This phenomenon is also shown by the Coinstone, a concretionary limestone horizon in the Sinemurian of Dorset, which has undergone a complex history of diagenetic $CaCO_3$ segregation to form nodules which were subsequently exhumed, encrusted, bored, reburied and pyritised, all within the time interval of an ammonite zone (Hallam 1969 *c*). Fürsich (1971) has made a thorough analysis of multiple hardgrounds with stromatolitic overgrowths in the Bajocian of Normandy and has demonstrated a complex process of reworking, excavation and sedimentary cavity-filling within this highly condensed sequence (fig. 4.1).

By relating his observations on hardgrounds in Middle Jurassic limestones in the eastern Paris Basin to the broader stratigraphic context and to recent discoveries in the Persian Gulf, Purser (1969) has made a significant contribution to our understanding of Jurassic limestone sequences.

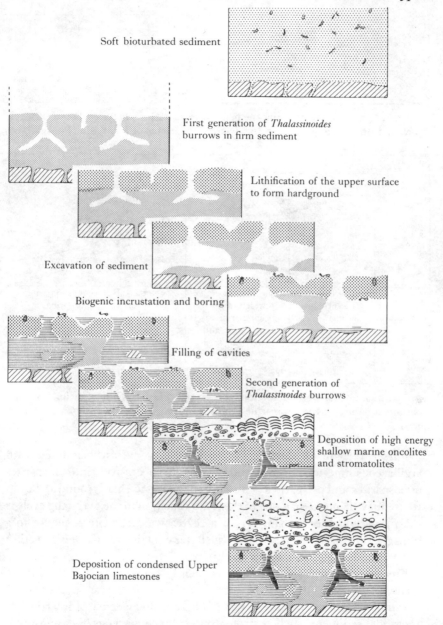

Soft bioturbated sediment

First generation of *Thalassinoides* burrows in firm sediment

Lithification of the upper surface to form hardground

Excavation of sediment

Biogenic incrustation and boring

Filling of cavities

Second generation of *Thalassinoides* burrows

Deposition of high energy shallow marine oncolites and stromatolites

Deposition of condensed Upper Bajocian limestones

FIGURE 4.1   Inferred succession of events in the deposition of the condensed Bajocian of Bayeux, Normandy. (Adapted from Fürsich 1971.)

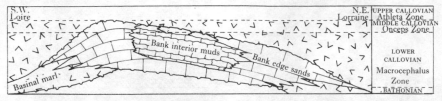

FIGURE 4.2 Section showing the general relationships between bored surfaces, rock and time units within the Callovian of the eastern Paris Basin (Purser 1969).

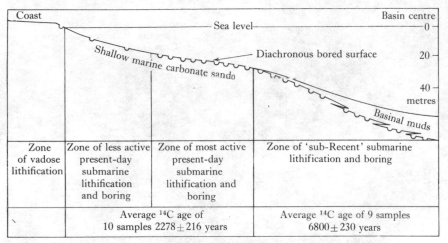

FIGURE 4.3 Generalised section across the Persian Gulf showing the diachronous nature of the submarine-lithified and bored Recent carbonates (Purser 1969).

Purser recognised three major sedimentary cycles, beginning with argillaceous limestone or marl and passing up successively into biosparites or oosparites and pelsparites, with a bored, encrusted hardground at the top (there may be more than one of these). Transition to the overlying cycle is sharp, and signifies a change from a regressive carbonate sequence to a transgressive marly sequence since the latter contains ammonites and other fossils signifying open marine conditions.

Emergence above sea level has frequently been invoked in the past to account for hardgrounds. That this is improbable in the case of the beds in question is signified by several facts. There is no evidence of intertidal or supratidal sediments such as stromatolites, bird's eye vugs or evaporites in the immediately subjacent deposits, and the hardgrounds are directly overlain by low energy, open marine deposits. The hardgrounds are also diachronous, in such a way that widespread contemporary emergence of a carbonate bank can be ruled out. Thus in Purser's Callovian example it can be demonstrated that the oldest hardground occurs on the basinal flanks of the carbonate bank of Bahamian type which is invoked as the likely depositional environment, while the youngest occurs on the crest (fig. 4.2).

Purser referred to the discovery in Recent calcareous deposits of the Persian Gulf of subaqueously cemented surfaces related to slow sedimentation. Evidently a low sedimentation rate is the only essential condition that need be invoked to account for such cementation. The pertinence of this inference to Jurassic limestone hardgrounds, which characterise condensed sequences, is obvious. A slowing down of carbonate sedimentation of the cycles was related by Purser either to a cooling climate or to a deepening sea. The plausibility of the latter can be demonstrated diagrammatically (fig. 4.3) by illustrating the effects of the Flandrian transgression on sedimentation in the Persian Gulf.

A striking feature of many hardgrounds is their remarkably planar surface extending over a considerable distance. This can be well seen, for instance, in quarries of Bathonian limestone near Cirencester in southwest England, where hardgrounds are exposed for hundreds of square metres. The phenomenon could be the result of either physical or organic processes or, perhaps most likely, a combination of the two. The prolonged to-and-fro movement of shell and sedimentary particles under the influence of currents could result eventually in marked planation. On the other hand, by analogy with processes known to operate on limestones in tropical intertidal zones, it could be that the phenomenon is largely due to organisms. The activities of rock borers weaken the surface layers of limestone to such an extent that physical erosion is greatly facilitated. Moreover, the activity of rock grazers and browsers, notably chitons and gastropods, is important in planing these surfaces.

The ecology of a Bathonian hardground fauna in Wiltshire is described in detail by Palmer and Fürsich (1974).

Marl–limestone sequences

A widespread facies associated with more purely carbonate sequences consists of regular small-scale alternations of microsparitic argillaceous limestones, containing about 75–85 per cent of $CaCO_3$, with marl. The old debate as to whether these alternations were primary, sedimentational or secondary, diagenetic features has been modified in recent years to the subtler question of assessing the relative importance of sedimentary and diagenetic processes.

Such deposits characterise, for example, the Hettangian and Lower Sinemurian Blue Lias formation of southwest England and South Wales (Hallam 1960, 1964); they are rich in benthonic fossils and were evidently deposited in quite shallow water. Thus in South Wales they pass laterally within a short distance into bioclastic and conglomeratic limestones which have transgressed over a local topographic 'high' formed by Carboniferous limestone (Wobber 1965). Comparable deposits in the Lower Pliensbachian Belemnite Marls formation of Dorset exhibit scour surfaces, another indicator of shallow water (Sellwood 1972 *b*).

The best evidence for a primary origin for both these formations is provided by the mottling of trace fossils such as *Chondrites* and *Diplocraterion*, which are produced by the piping down of darker marl into lighter, more calcareous sediment and vice versa. Strictly speaking, however, this phenomenon could have been preceded by some very early chemical segregation of $CaCO_3$ directly beneath the sediment–water interface, while the sediment was still accessible to burrowers a short distance above. In the case of the Blue Lias there is evidence that rhythmic segregation in diagenesis has subsequently enhanced the sedimentary contrasts brought out by the burrowing (Hallam 1964).

In a more recent study of similar facies in the Lias of Portugal the conclusion was reached that diagenesis was the dominant factor in producing the limestone–marl cyclic sequence (Hallam 1971 *c*). Since some of the arguments may have more general application they are worth noting here.

(i) No correlation is apparent between the cycles and any other sedimentary or faunal feature, as might be expected if the control related to the depositional environment; thus shell bands are distributed randomly with respect to the rhythms.

(ii) The regular limestone bands may be seen to pass into bands of limestone nodules which are generally acceptable as secondary. The microtextures of all the limestones give no indication of primary deposition but are perfectly consistent with interstitial crystallisation during diagenesis.

(iii) Although the thickness of the total succession varies considerably in different localities the thickness of the limestone bands remains essentially constant and the number of such bands declines in approximate proportion to the thinning of the total sequence. The modal thickness and range of thickness for the Portuguese limestones are in fact virtually identical to those for the Blue Lias of Dorset and Glamorgan (fig. 4.4). In other words, rate of sedimentation appears to be irrelevant to the formation of the cycles.

(iv) The various indications that exist of tectonic and sedimentational stability during deposition tend to render implausible the invocation of frequent environmental oscillations of the order of several hundred per stratigraphic stage, such as are demanded by the thicker sequences.

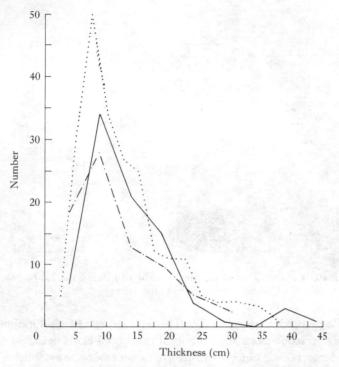

FIGURE 4.4 Plot of limestone thickness against number for the Lower Pliensbach-ian of Portugal and the Hettangian and Lower Sinemurian of southwest England and South Wales. ...., Blue Lias, Glamorgan; –·–·–, Blue Lias, Dorset; ————, Late Pliensbachian, San Pedro de Muel, Portugal (Hallam 1971 c).

(v) Whereas in thick sequences such lithologies as shales, sandstones and coarse-grained limestones frequently persist for several metres or more, this never seems to be the case in marly sequences, which always seem to consist of thin, regular alternations of more or less calcareous beds in which the only evident variable is $CaCO_3$ content. It seems somewhat specious to argue that, whereas the limestones as we observe them today are essentially diagenetic in origin, the horizons of $CaCO_3$ segregation were nevertheless subtly determined by some environmental factor that has otherwise left no trace.

Fürsich (1973) has related the thin-bedded nodular limestones found in parts of the English Upper Oxfordian to *Thalassinoides* and suggested a control of nodule formation by the activity of burrowing crustaceans (fig. 4.5). He argued that the large amount of organic matter in the burrows,

71

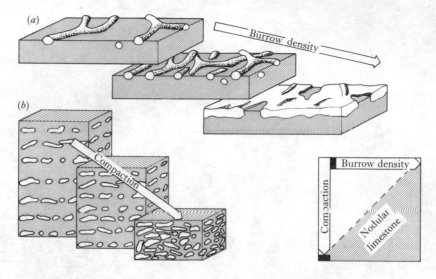

FIGURE 4.5   Two mechanisms leading to the formation of
nodular limestones (Fürsich 1973).

both in the mucus lining and in faecal pellets, served as an inducement for
$CaCO_3$ precipitation because of the local rise in pH engendered when the
material decomposed. Once a diffusion gradient became established carbonate precipitation continued and layers of concretions grew.

It must be admitted that at present the process of rhythmic segregation
which we seem obliged to infer is only poorly understood. It will be
considered further in the next chapter, when the origin of nodular
limestones in the Mediterranean region is discussed.

Palaeoecological aspects

On the whole, the faunas, both of body and trace fossils, differ little from
those of the terrigenous clastic and ferruginous facies, there being little
simple correlation between particular types of sediment and fossil. The
important distinction once more seems to be between shallower and deeper
water associations, as determined by a variety of criteria (fig. 4.6). Thus the
shallower water deposits, characterised by relatively condensed sequences
with mixed algal limestones, microsparites, biosparites, pelsparites and
oosparites, and the presence, locally, of cross bedding and erosion surfaces, are richer in large, suspension-feeding bivalves and the trace fossils
Diplocraterion, *Rhizocorallium* and *Thalassinoides*. The deeper water deposits, characteristically deposited in basins rather than on banks or
swells, tend to be uniformly fine-grained and more argillaceous. The above

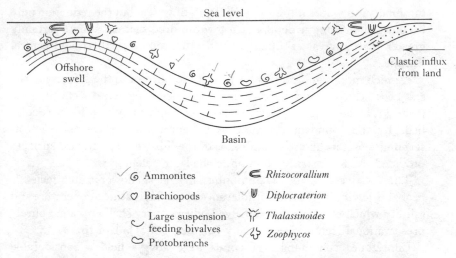

FIGURE 4.6 Schematic portrayed of salient faunal characteristics in relation to local variations of water depth (Hallam 1972 *a*).

mentioned fossils are much rarer but ammonites much more frequent. Brachiopods and echinoderms, in fact most benthos, tend to be richer in the shallower water facies. In general, reef-building corals, calcareous sponges, bryozoans and brachiopods are commoner than in the corresponding shallow water deposits of the terrigenous clastic and ferruginous facies.

The characteristic, in fact almost the only, trace fossils of the deeper water facies are *Chondrites* and *Zoophycos*, although both occur also in the shallower deposits. The highly distinctive burrowing structures of *Zoophycos* cover bedding planes in a vast number of fine-grained marly calcareous deposits in Spain, Portugal and southern France. Many of these so-called *Cancellophycus* limestones seem to be almost barren of any other macrofossils but ammonites.

The sparsity of benthos in the deeper water calcareous facies, as opposed to the corresponding shaly deposits of the terrigenous clastic facies, calls for comment. It cannot merely be a matter of reduced food supply at depth, since these deposits pass laterally over short distances into, or alternate at brief intervals in the stratigraphic succession with, deposits which could not have been laid down in more than a few metres of water. Furthermore, there are no indications, in the form of submarine breccias or turbidites, of strong topographic contrasts on the sea bed. Moreover, some of these deposits exhibit scour surfaces, recorded for instance by Sellwood (1972 *b*) from the Belemnite Marls of Dorset. It is true that the phytoplankton food supply does diminish with depth of water, but the reduction only becomes significant at depths of a few hundred metres, almost certainly greater than any values attained in the Jurassic of northern and central Europe.

A much more significant factor is likely to have been the softness of the sea bed, which would have caused suspension feeders to choke quite readily (Hallam 1972*a*, Sellwood 1972*b*). This is confirmed for particular beds by the abundant colonisation by encrusting organisms, such as serpulids, oysters and foraminiferans of the limited areas of hard surfaces provided by, for instance, ammonite shells or belemnites.

The relative sparsity of ammonites in the shallower water facies is probably attributable to their intolerance of the fluctuating temperature and salinity which characterise extensive bodies of very shallow water; purely preservational factors can generally be ruled out (Hallam 1972*a*).

Limestones containing reef corals have always held a considerable fascination for the Jurassic palaeoecologist. Those in the Upper Oxfordian are by far the most familiar, but the best exposed coral reefs in northwest Europe are probably to be found today in the Lower Bajocian (Humphriesianum Zone) and are beautifully displayed in a large abandoned quarry at Malancourt, a few kilometres northwest of Metz in the eastern Paris Basin (Hallam in press).

The reefs are regularly spaced ovoid masses of unbedded or crudely bedded limestone in the midst of well-bedded bioclastic limestones and marls, and are up to about 20 m in height and 30 m in breadth. The dominant corals are *Isastrea* and *Thamnasteria*, but branching corals are also present. They occur primarily as irregular flat or curved sheets of massive saccharoidal limestone, which are generally parallel or almost parallel to the bedding and are frequently seen to branch. Shelly detrital limestone occurs between the sheets. The thicker sheets often show drusy cavities lined by large calcite crystals. Unlike in present day reefs, obvious algal structures appear to be absent. The associated invertebrate macrofauna is both abundant and diverse; it is dominated by strongly ribbed *Chlamys* and other large pectinids and limids, *Lopha*, brachiopods, cideroids and the gastropod *Bourguetia* – a fauna strongly resembling that associated with the very similar Upper Oxfordian patch reefs, as described for instance by Arkell (1935), and the Lower Kimmeridgian patch reefs exposed on the coast directly south of La Rochelle, France. Neither the Oxfordian nor Kimmeridgian coral masses (which, it can be argued, are too modest in size to qualify as reefs) exhibit the laminated structure of the Malancourt examples.

Analysis of the coralline structures and the enveloping sediments suggests that the reefs grew as low mounds on a very shallow sea floor, at geographic intervals of a few tens of metres. There is no evidence of original slopes of high gradient, nor of contemporary erosion and reworking, and those scattered corals that occur in the intervening bioclastic limestones appear to be *in situ*. It appears, therefore, that emergence of the reefs into a zone of intensive surf action can be positively ruled out.

Another important group of colonial animals associated with limestones are the sponges, which are important at certain horizons. Thus there is a widespread development of spongiferous deposits in the Upper Bajocian, extending northwards from northern Spain into Normandy and Dorset, and another one in the Upper Oxfordian, extending from eastern Spain into the Jura Mountains and Swabia. Reefoid structures are claimed to occur in Swabia, where the facies ranges into the Lower Kimmeridgian (Geyer and Gwinner 1968) but more generally one can speak only of an abundance of individual sponge colonies measurable in cms or dms rather than metres. Compared with the coraliferous deposits noted above, the sponge beds are relatively deficient in bivalves but rich in brachiopods and ammonites, suggesting a somewhat deeper water facies.

A particular type of fauna more characteristic of the calcareous than terrigenous clastic facies is that of rock borers, which may penetrate either hardgrounds, derived pebbles or calcareous shells. Besides the probable polychaetes described by Hölder and Hollman (1969) these may include clionid sponges, cirripede crustaceans, bivalves and algae, as described for instance from the Lower Liassic Oolithenbank and Kupferfelsbank condensed limestones of Baden–Württemberg (Schloz 1968, 1972; see also fig. 4.7).

Marginal marine environments within the calcareous facies tend to be characterised by the presence (or former presence) of evaporites, suggesting hypersaline lagoons. Whereas faunal diversity is strongly reduced, with the substantial exclusion of stenohaline groups such as corals, cephalopods and many other molluscs, brachiopods, bryozoans and echinoderms, those few species that were able to tolerate the environment, mainly of bivalves and gastropods, may occur in huge quantities. They may cover whole bedding surfaces, in sequences that are not condensed, so that the factor of abnormally slow sedimentation can be excluded. Examples are parts of the Purbeck Beds of Dorset and the Coimbra Beds near the base of the Lias in Portugal. There is no reason to presume a greatly increased food supply in hypersaline lagoons, so the explanation of the high population density should be sought elsewhere. It has been explained tentatively, with reference to modern ecological concepts, by postulating so-called *r*-selection, for fecundity, which ensures effectively the perpetuation of the

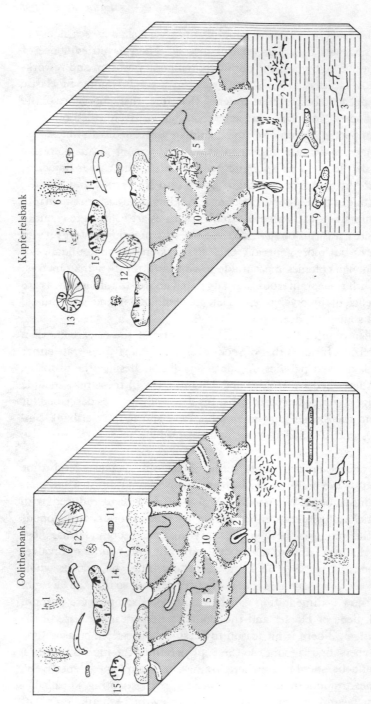

FIGURE 4.7 Trace fossils, both burrows and borings, in Lower Liassic condensed limestones in Baden–Württemberg, Germany, as seen in vertical section and on undersurfaces: 1, bioturbation textures; 2, *Chondrites*; 3,4, pyritised worm burrows; 5, possible worm trails; 6, dwelling burrows; 7, *Pholadomya* burrow; 8, *Rhizocorallium*; 9, *Spongeliomorpha*; 10, *Thalassinoides*; 11, algal borings; 12, clionid borings in shells; 13, cirripede borings in shells; 14, indet. borings in *Cardinia* shells; 15, ? *Simonizapfes*) in *Gryphaea* shells; 14, indet. borings in *Cardinia* shells; 15, ? lithophagid borings in limestone pebbles (Schloz 1968).

species in unstable environments where high mortality is inevitable (Hallam 1972*a*).

A final point that ought to be made here concerns the occasional discovery of micritic faecal pellets within 'protected' areas of fine-grained limestones, such as the interior of shells, where they are clearly recognisable in thin section against a background of sparry calcite cement (e.g. Hallam 1971*c*: fig. 7). Without such protection, the soft pellets eventually compact together to produce a micritic rock which shows little or no hint of its pelletal origin. We may infer that faecal pellets were a much more important component of many fine-grained Jurassic rocks than present textures signify.

EXAMPLES OF FACIES ANALYSIS

Great Oolite (*Bathonian*) of southern England

The first systematic attempt to compare British Middle Jurassic limestones with deposits in the Bahamas has been undertaken for the Great Oolite group of the Bath area of Somerset by Green and Donovan (1969). They subdivided the succession into four formations, the Combe Down Oolite and the Twinhoe Beds, which correlate with the upper part of the Fuller's Earth Clay further south, and the Bath Oolite and Upper Rags, which correlate with the Forest Marble in the same direction. In terms of the regional picture, the more or less clay-free Bath area lies on the so-called Cotswold Shelf, and the clays to the south were deposited in the deeper water of the Dorset Basin (Martin 1967).

The thicker oolite sequences are thought to represent so-called 'oolite deltas' like those which have been described on the Trucial coast, where they are related to patterns of tidal currents and have a maximum depth of 2 m. The thinner oolite zones could represent areas of tidal channelling. The Twinhoe Beds, including the well-known ironshot limestone, are finer-grained and richer in shells; they probably signify slightly deeper, quieter water. The Bath Oolite contains thick cross-bedded sets and must signify the former presence of mobile oolite banks. The highest formation, the coraliferous Upper Rags, signifies the development of patch reefs when the sea bed became stabilised. The absence of more widely distributed coral beds is attributed to the extensive area of very shallow sea to the south, which limited free renewal of nutrient-rich water from the open ocean. (It is well-known that in Recent environments the most prolific coral growth takes place on the edge of banks, probably for this reason.)

In Oxfordshire, to the northeast of the area considered by Green and

Donovan, the lower part of the Great Oolite is represented successively by the Chipping Norton, Sharp's Hill and Taynton formations, whose environments of deposition have been interpreted by Sellwood and McKerrow (1974). The Chipping Norton Formation is composed of grain-supported limestones, signifying a high energy environment, which were deposited on a swell located on the site of the Bajocian Moreton in Marsh Swell. In a small basin directly to the east an influx of sand can be recognised, which entered the area from the east and appears to represent deltaic sediments derived from the London Platform. In another basin to the west of the swell limestones interdigitate with clays of the Lower Fuller's Earth Formation.

The clays of the Sharp's Hill Formation contain a diverse marine benthos, including bivalves, gastropods, corals, echinoids and rhyncho-nellid brachiopods, and appear to signify a slight deepening of the shallow sea, which also transgressed over the source area of the underlying sands. Locally around Stonesfield in the eastern basin and Stow-on-the-Wold in the western basin, this transgression is marked by silty limestones of the Stonesfield Member, which is thought to have accumulated on topo-graphic lows as condensed mixtures of redistributed terrestrial and marine faunas and sediments. Locally, fresh water gastropods of the genus *Viviparus*, ostracods and charophytes were washed into the Sharp's Hill clays from the land. In some places the early phase of the marine transgression resulted in non-deposition and the development of hard-grounds capping the Chipping Norton Formation.

The three upper formations of the Great Oolite in the same region are, from bottom to top, the White Limestone, Forest Marble and Cornbrash. Klein (1965) described small channels in the lower two formations which he interpreted as signifying lateral erosion and deposition by tidal currents in the intertidal zone. Subsequently, a palaeoecological analysis by McKerrow *et al.* (1969) was held to confirm an intertidal origin for the White Limestone, but the Forest Marble was thought to be shallow sub-tidal. Bearing in mind the work of Reineck and others in the German Bay, small-scale channelling is not in itself conclusive proof of intertidal as opposed to shallow subtidal conditions, and the faunal data are somewhat equivocal. More convincing evidence of emergence has been put forward by Palmer and Jenkyns (1975).

Locally at the top of the White Limestone, in a zone near Oxford, a thin stromatolitic bird's eye limestone has been discovered, which indicates elevation into the high intertidal or supratidal regime. Below this the beds consist of micritic limestones with a marine fauna. Above it, to the southwest, the Forest Marble is a grain-supported, partly oolitic limestone

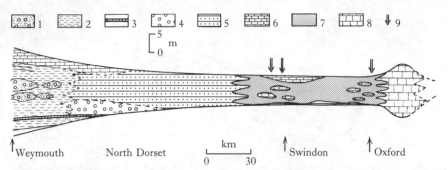

FIGURE 4.8 Facies relationships within the Osmington Oolite (Upper Oxfordian) Group of southern England. 1, quartzose oosparites etc.; 2, marls and nodular limestones; 3, pisolite; 4, cross-bedded oosparites; 5, oosparites, intrasparites; 6, *Rhaxella* biomicrites; 7, 8 coral limestones; 9, clay-filled channels (Wilson 1968 b).

with reworked shells, signifying an open, current-swept shelf. To the northeast there is more clay in the Forest Marble, with finer-grained limestone and more fragments of lignite. The fauna has brackish and fresh water elements including *Viviparus*.

The coincidence of facies change in the Forest Marble with the zone of bird's eye limestone suggests the emergence of a barrier island or spit, isolating a lagoon to the northeast into which clastic sediments were brought from the land. Subsequently the end-Bathonian transgression, signified by the Lower Cornbrash, swamped the whole area, and the facies once more became laterally homogeneous, in the form of slowly-deposited non-oolitic shell-fragment limestone.

Upper Oxfordian of southern England

It was in the Corallian beds of Oxfordshire that Arkell (1935) did his pioneer environmental study of coral patch reefs. More recently Wilson (1968 a) has attempted a palaeogeographic synthesis of the Corallian in southern England. He subdivided the depositional area into three major units (which, incidentally, can be recognised through a greater time span of the Upper Jurassic). These are the Wessex and Wealden basins, separated by the Portsdown Swell. Clay facies is dominant in the Wealden Basin, but quartz sands increase proportionally westwards. The source of the sands of the Lower and Upper Calcareous Grits is considered to lie to the north, and probably includes the London Ridge. It is with the limestone formations, however, that Wilson was principally concerned, and another paper (1968 b, see figs. 4.8 and 4.9) is devoted to a facies analysis of the Osmington Oolite Formation.

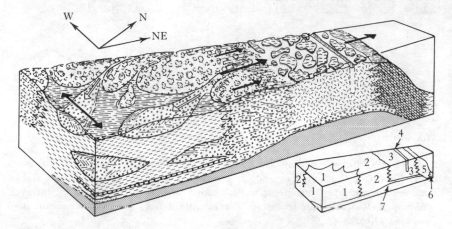

FIGURE 4.9   Facies model of the Osmington Oolite Group. 1, true oolite facies; 2, oolite freestone facies; 3, coral limestone facies; 4, channels within 3; 5, Wheatley limestone facies; 6, argillaceous facies; 7, quartz sands (Wilson 1968 *b*).

Five facies were distinguished:
   1. True oolite
   2. Oolite freestone
   3. Coral limestone
   4. Wheatley limestone
   5. 'Phyllosilicate clay' (unfortunately the types of clay mineral were not determined).

The true oolite facies exhibits a repetition of coarsening upward units, in the sequence: beds with small intraclasts, 'quiet water' oolites, normal oolites, beds with algal or skeletal debris. The presence of well-developed ooids suggests depths of less than 3 m. Since the sequence suggests progressive shallowing into a more turbulent environment it is thought to relate to tidal flat regression, by analogy with recent deposits. The ubiquity of spicules of the sponge *Rhaxella* suggests that perhaps the organisms acted as sediment stabilisers of the bioturbated oolite beds, in contrast to the cross-bedded strata of the oolite freestone which indicate the original presence of mobile oolite shoals. By analogy with the Great Bahama Bank, the sponges would have acted in the way that sea grasses and green algae do today, as sediment baffles. Measurements on the cross bedding indicate dominantly east–west, presumably tidal flow.

The coral limestone facies represents patch reefs, and the Wheatley limestone facies largely micritised skeletal debris swept northwards from the reef zone. The clay facies represents the 'background' sedimentation and locally fills inter-reef channels. The general facies pattern was influenced by the morphology of the underlying banks of quartz sand. Where these are thickest, between Swindon and Oxford, the highest energy carbonates (coral biolithites) accumulated.

The preservation of the corals has been investigated by Talbot (1972). Naturally, all the original aragonite has been replaced. The growth of drusy acicular aragonite within the corals and precipitation of micritic cement, both of which may have occurred during the life of the reefs, has considerably influenced the end product. Trapped sediment may have been vital in providing nuclei around which calcite could precipitate. When corals lost $CaCO_3$ on a large scale, the structure was only saved from collapse by the early lithification of interskeletal cement.

Talbot (1973) has also attempted an environmental interpretation of the relationship of the limestone horizons within the Corallian to the inter-bedded terrigenous clastic sediments. Arkell (1933) recognised three sedimentary cycles composed successively of clay, sandstone and limestone. Talbot's analysis leads to a different scheme of cycles, based upon his understanding of the respective depositional environments.

Six different environments are recognised:

1. *Offshore shelf*. The carbonates of this environment are mainly molluscan biosparites with variable quantities of ooids, fish teeth and scales and phosphatic nodules. The highly bioturbated beds contain a diverse fauna composed principally of bivalves together with the alga *Girvanella*. An open marine, shallow subtidal environment of slow deposition is implied. The associated clays are also strongly bioturbated.

2. *Reef*. This environment is as described above for the Osmington Oolite Group.

3. *Nearshore subtidal*. This consists of clays, silts and sand which are usually bioturbated but also have beds exhibiting ripple lamination and wavy bedding, with a bivalve fauna. Comparison is made with the Recent sediments of the German Bay of the North Sea.

4. *Intertidal and estuarine*. The carbonates are oosparites with planar cross bedding, sometimes with thin layers of clay draping over the foreset laminae. The sparse fauna appears to have been washed in by storm action, and a comparison is made with Bahamian ooid banks. The clastic deposits consist of fine to medium sands exhibiting planar cross bedding, flat and ripple lamination, flaser bedding and clay drapes; small channels occur. Shells are rare but trace fossils of the ichnogenera *Ophiomorpha*, *Diplocraterion* and *Arenicolites* abundant. The Bencliff Grit east of Osmington

Mills on the Dorset coast contains abundant plant material and is thought to be an estuarine rather than a tidal flat deposit, as Wilson interpreted it.

5. *Lagoon.* This facies consists in its carbonate portion of highly bioturbated oo- and biomicrites containing vast numbers of the rhaxes of *Rhaxella* and partly micritised molluscan shells; *Teichichnus* is the dominant trace fossil. The non-carbonate portion is composed of sideritic clays and silts with chamosite and limonite. The high density, low diversity fauna is dominated by the oyster *Deltoideum delta.*

6. *Supratidal.* This is a minor environment, indicated by clays containing rootlets, thought to signify tidal marsh deposits.

Fig. 4.10 illustrates the sequence of facies and formations as interpreted by Talbot. An important part of his interpretation concerns the recognition of four major erosion surfaces in the areas north of Dorset, which truncate the underlying beds and are marked by layers of pebbles and phosphatic nodules. Lithological change to the overlying beds is sharp and the surfaces can reasonably be held to define four sedimentary cycles. Two of these cycles consists, in upward sequence, of limestone, clay and sand, one of clay and sand, and one (the Osmington Oolite Group) essentially of limestone. The Ringstead Coral Bed, normally placed within the Corallian, is held to define the base of a succeeding Kimmeridge Clay cycle.

In Talbot's very reasonable interpretation, the cycles mark periods of regressive sedimentation between each pair of erosion surfaces, signifying shallowing of the shelf and its replacement by a shoreline complex of beaches and lagoons (fig. 4.11). It is thought that when sea level remained constant for a period the accumulation of sediments outstripped subsidence and led to lateral seaward migration of the clastic material. A sharp rise of sea level caused a strong reduction in terrigenous sedimentation at the same site and carbonate sediments were deposited slowly instead, so commencing a new cycle. The Osmington Oolite cycle appears to mark a more notable rise of sea level and transgression which caused the terrigenous clastic zone of deposition to be pushed back out of the region.

The cycles do not appear to be markedly, if at all, diachronous, and indeed there is stratigraphic evidence suggesting that the major Osmington Oolite transgression correlates with a similar event within the Antecedens Subzone of the southern French Jura, as marked by overstep of the Birmensdorf Beds (Enay 1966). The same event appears also to find expression in the Swiss Jura (Gygi 1968).

Interesting work has also been done on the cementation of the Corallian limestones. Both Talbot (1971) and Davies (1971) recognised an early stage of non-ferroan calcite cement followed by quite distinct ferroan calcite

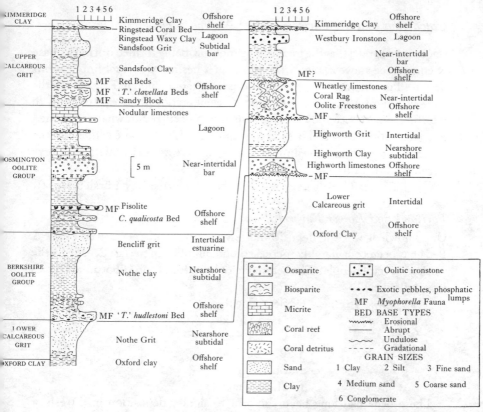

FIGURE 4.10 Facies interpretation of the 'Corallian' Sequence
(Upper Oxfordian) of southern England (Talbot 1973).

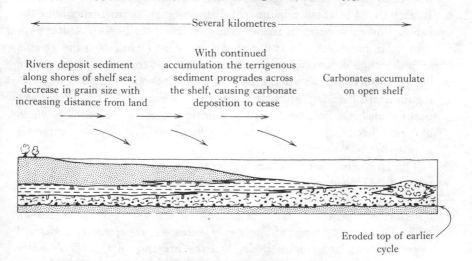

FIGURE 4.11 Suggested origin of the limestone–clay–sand cycle in the
southern English Corallian (Talbot 1973).

stages, and agreed that the former was most probably formed during a phase of emergence above sea level. If this is so it has more general implications because Purser (1969) recognised the same phenomenon in French Middle Jurassic limestones. The only type of cementation that he attributes to intertidal emergence, however, is that which occurred in what are termed microstalactitic druses. These are asymmetrical, with more cement on the lower than upper side of sediment grains, and are associated with bird's eye vugs and stromatolites.

Corallian trace fossils have recently been described by Fürsich (1974).

Portland and Lower Purbeck Beds of Dorset
Our knowledge of the depositional envirionment of the Lower Purbeck Beds has been considerably increased and modified in the last decade by the work of several people. The sequence of limestones and subordinate marls is as follows:

6. Soft Cockle
5. Hard Cockle
4. Cypris Freestones
3. Broken Beds
2. Soft Cap
1. Hard Cap

These beds had long been thought to signify deposition in a fresh or brackish water lagoon, although the Broken Beds had been widely considered to signify collapse following the solution of gypsum or anhydrite. In fact traces of these calcium sulphate minerals occur throughout the Lower Purbecks, either at the surface as calcite pseudomorphs after gypsum, or in boreholes as nodular anhydrite. West (1964, 1965) has given a very thorough description of the diagenetic stages to which the beds have been subjected and recognised the presence of celestite and other interesting minerals. Pseudomorphs after halite are common in both the Hard and Soft Cockle beds. Precipitation of evaporites suggests hypersalinity, which is contradicted neither by the algae (Pugh 1968) nor by the ostracods (Anderson, in Wilson *et al.* 1958).

The Hard and Soft Caps are evidently composed largely of sediment trapped by blue–green algae (Brown 1963, Pugh 1968). The other Lower Purbeck limestones are either micrites or pelmicrites, with layers of obvious faecal pellets (probably crustacean) in the Soft Cockle beds (Brown 1964). From the evidence of a westerly increase in the quantity of detrital quartz in the Hard Cockle Beds, and the presence of a brackish water molluscan fauna in the same beds only to the west of Mupe Bay, Brown inferred a coastline to the west, with the hypersaline lagoon locally being diluted by influx of fresh water. Florida Bay would seem to provide a reasonably close analogue at the present day. The regular limestone–marl cyclicity in the Lower Purbecks was considered by Brown to be primary in origin, but he gave no reasons for excluding the possibility of rhythmic unmixing of $CaCO_3$ during diagenesis.

Geochemical analysis of stable carbon isotope ratios supports the notion of strong fluctuations in salinity, but indicates a general decline of salinity from the Purbeck Beds into the Wealden. The basal Purbecks and the Cinder Bed give the highest values, which accords with the data from fossils and evaporites (Allen and Keith 1965).

The most pertinent analogy with modern deposits is that made by Shearman (1966) following examination of the basal Purbeck sediment core of the Warlingham borehole south of London. Shearman recognised sedimentary cycles of a type found on supratidal sabkha flats of the Trucial coast:

4. Carbonate mud: Lagoonal

Erosion surface, with intraformational conglomerate

   3. Nodular anhydrite in carbonate mud: Supratidal

   2. Stromatolitic laminae: Intertidal

   1. Carbonate mud: Lagoonal

The relevance of this interpretation to the Hard and Soft Caps of Dorset is obvious.

Like the Purbeck Beds, the Portland Beds are best developed in Dorset but extend northwards into Wiltshire, Oxfordshire and Buckinghamshire; they occur also in the Bas Boulonnais and are known from subsurface exploration in Sussex and Kent. All these regions have been studied recently by Townson (1971) but, not unnaturally, most attention is paid to Dorset. Townson divided the Portland *Group* into a lower, Portland Sand, and an upper, Portland Limestone, *Formation*.

The Portland Sand marks an upward continuation of the sandy and silty Upper Kimmeridge Clay. The beds contain dolomite, which becomes the dominant mineral towards the top of the formation. The origin of this dolomite poses a problem since Recent dolomite is described mainly from the supratidal environment and in Townson's facies model the Portland Sand was deposited in deeper water than the marine Portland Limestone.

In his tentative explanation, Townson applied the first phases of the model invoked by Schmalz (1969) to account for deep water salt deposits (if indeed such deposits exist), which involves a reflux of magnesium-rich sea water. Townson noted the anomalous absence of dolomite from the Lower Purbeck beds. Since elsewhere in his thesis he gave evidence which suggests that the Portland–Purbeck transition is diachronous in southern

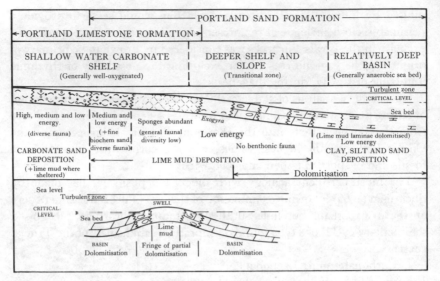

FIGURE 4.12 General model for environments of deposition of the Portland Group of Dorset. The lower diagram shows the influence of the swell on depths of water and hence on dolomitisation. Vertical scale exaggerated (Townson 1971).

England, it is not unreasonable to postulate a Purbeck-type supratidal/ lagoonal environment in an onshore direction during deposition of the Portland Sand (fig. 4.12). In the model proposed, following precipitation of $CaSO_4$ in the 'Purbeck' environment, magnesium-enriched water flowed off the shallow carbonate bank into a deeper stagnant basinal zone. Below a certain critical level the oxygen content of the interstitial water was sufficiently low, and the pH sufficiently high, to allow the dolomitisation of $CaCO_3$.

The Portland Limestone consists of sponge-rich cherty micrites passing up into higher energy biocalcarenites and oolites (the classical building freestone). Within the Dorset area a swell separated western and eastern basins and influenced the depth and type of sedimentation to some extent. The fauna consisted mainly of molluscs. The rarity of brachiopods, corals and echinoids and the virtual absence of belemnites and crinoids were thought to be due, at least partly, to fluctuating salinities on a shallow bank, with periodic tendencies towards hypersalinity. Epifaunal bivalves were largely excluded from soft lime muds in the quieter water areas, and from mobile carbonate sands now represented by cross-bedded oolites.

The idealised model of Portland and Purbeck-type environments por- trayed in fig. 4.12 is not intended as a scale representation of the situation in Dorset at any one time, but shows, in exaggerated fashion, how the facies is thought to have related to the depth and energy level of water and to the presence or absence of a swell. In the deepest water zone laminated lime mud was laid down and subsequently dolomitised. In the zone shallower than this the sea bed was soft and epifaunal bivalves were only present where a local hard substrate, such as ammonite shells, was available. In the shallowest marine zone ooids were able to form, and a fauna existed where the bed was sufficiently stabilised. Thus carbonate sand bodies, when they became inactive, were the site of colonisation by both epifaunal and infaunal bivalves and burrowing crustaceans. Locally on the swells, as well-displayed in the Isle of Portland exposures, patch reefs of oysters, bryozoans and the red alga *Solenopora* grew to cover areas of several square metres. In slightly deeper water and in the shallow protected water behind the carbonate sand bodies, sponges flourished, as signified by the abundant spicules of *Rhaxella* and, to a lesser extent, *Pachastrella*. Margining the shallow sea, the growth of stromatolitic algal mounds served to partly isolate the hypersaline lagoon where Purbeck-type deposits were laid down.

Townson's environmental model is an attractive example of Walther's celebrated 'law' of facies. The succession of facies from offshore to onshore corresponds with the vertical regressive sequence from the top of the Kimmeridge Clay to well into the Lower Purbeck beds.

High Upper Jurassic of West Germany

The Lower Tithonian Solnhofen lithographic limestone of Bavaria is even more famous for its excellently preserved fossils than are the Swabian Posidonienschiefer. It consists of well-bedded micritic layers of almost pure $CaCO_3$. Argument has continued over many years on the origin of the micrite and the character of the depositional area, in particular how the unusual preservation of fossils has been accomplished. Barthel (1970, 1972) and Van Straaten (1971) have recently turned their attention to these problems.

Characteristically, the limestone consists of thin-bedded micrite, almost pure $CaCO_3$, with individual beds ranging up to 30 cm and separated by marl partings. Contrary to the impression one gains from perusing museum collections, it is rather poor in fossils, though a large variety has been collected over the years from the celebrated quarries. The only common macrofossil is the pelagic crinoid *Saccocoma*. Ammonites are not uncommon and the same is true of fishes, of which about 150 species have been recognised. Arthropods and dibranchiate cephalopods are rare, and jellyfish, reptiles and *Archaeopteryx* very rare.

Electron microscopic study has revealed the presence of coccoliths as an important component of the limestone, but some material may have been

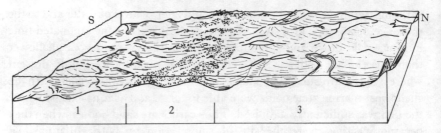

FIGURE 4.13 Reconstruction of the sea floor at the start of the Tithonian in Bavaria. Uplift in north, downward tilt in south. 1, open sea with relief of sponge reefs; 2, uplifted dead sponge reefs with corals growing on their tops; 3, dead sponge reefs in the backreef lagoon. Lime mud, now Solnhofen Limestone, was deposited in basins fringed by these reefs (Barthel 1970).

derived from other sources, either as fine organic detritals or as inorganic precipitates. The perfect preservation of fossils has in the past been explained by the episodic withdrawal of the sea from a shallow lagoon. Animal remains were left stranded and dessicated and were subsequently covered by sediment as the sea returned. Barthel opposed this interpretation and argued in favour of a continuous cover of sea water.

To appreciate properly Barthel's environmental model (fig. 4.13) it is necessary to consider the Solnhofen Limestone in its broader stratigraphic context. By the end of Lower Kimmeridgian times extensive growth of sponge–algal reefs had built up a complex relief on the sea floor of the Suabo-Bavarian region. (For detailed accounts of the celebrated sponge reef and inter-reef facies of Bavaria see Geyer and Gwinner (1968) and Hiller (1964).) Uplift exposed sponge reefs in the north but further south corals were able to grow, while still further south, in deeper water, sponge reefs continued in existence. Thus an approximately east–west girdle of coral and hydrozoan patch reefs developed along the line which now marks the southern margin of the Franconian Alb and the course of the Danube.

Hence a barrier came into existence to protect a backreef lagoon to the north from the agitated water of the open sea further south. The floor of this lagoon had a strong relief of dead sponge reefs, between which micritic sediment was deposited. Barthel estimated the water depth to be approximately 30 to 60 m. Continued uplift in the north forced the coral reefs to move progressively southwards, so that regression took place later in the Tithonian. Some allodapic (i.e. turbiditic) limestones occur in the east and horizons of slumping are more widespread, as are groove casts. These all testify to appreciable differential relief in the depositional area.

As evidence against periodic emergence, Barthel cited the absence of evaporites, stromatolites, bird's eye vugs and tidal channels. Furthermore, the sedimentation was remarkably uniform, with individual beds being traceable over several kilometres. It is very likely that the lagoon exhibited a tendency towards hypersalinity because the climate, as signified by the coral and reptile fauna, must have been warm. Density stratification was set up so that the bottom waters tended towards stagnation, which accounts for the almost complete absence of an indigenous benthonic life. Van Straaten supported the notion of oxygen deficiency, if not complete stagnation, in the bottom waters and suggested that the sediment had been reworked by storms and carried in suspension from a shelf area to the north.

Laboratory experiments undertaken by Barthel gave results consistent with subaqueous preservation of the abundant arthropods and the subordinate insects, jellyfish, pterosaurs and birds. Of the wide variety of crawling arthropods present, only *Mesolimulus* has left trails. This suggests that bottom conditions were lethal for all but the king crabs. Our knowledge of their living relatives suggests that they are indeed highly tolerant of varied salinity, temperature and oxygen content. The other arthropods were evidently washed in.

The experiments support field observations in suggesting that insects died on the surface waters of the lagoon and finally settled on the bottom in the position of death and according to the body shape. Thus butterflies and dragonflies mostly reach the bottom with their wings spread, while flies and grasshoppers, with wings folded, roll sideways once they touch the bottom.

A series of peculiar markings on bedding planes in the Solnhofen Limestone were interpreted in the past as the tracks of tetrapods or of fishes swimming in groups over the mud bottom. They have in fact been shown by Seilacher (1963) to be the drag marks of empty ammonite shells, produced by their tumbling and rolling over the soft mud. Rothpletz's (1910) well-known example of a venter impression beside a flat-lying ammonite impression is only one of a wide variety of markings attributable to this. Seilacher inferred that the ammonite shells were bowled along by an almost laminar current flow.

Barthel (1969) has also attempted an environmental interpretation of the early Upper Tithonian Neuberg Formation of Bavaria. The limestones include micrites, biomicrites and pelmicrites, exhibiting lamination, scour and fill, and burrowing structures, such as *Rhizocorallium*. As opposed to these evidently subtidal deposits, certain thin intercalations of calcirudite are thought possibly to be intertidal. The fauna, which is not of Tethyan type, is dominated by bivalves and gastropods which, judging by their closest living relatives, were more or less euryhaline. The presence of the

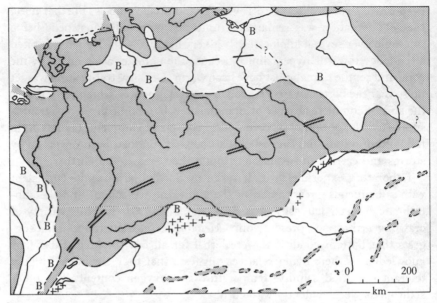

FIGURE 4.14   Palaeogeography of central Europe early in the Upper Tithonian.
B, brackish water zones; +, reefs; = =, zone of uplift; —, zone of subsidence.
Stippled areas are land (Barthel 1969).

foraminiferan *Anchispirocyclina* also suggests conditions of variable
salinity, which could in addition account for the absence of ammonites
and bryozoans and the occurrence of corals, brachiopods and echinoderms
only as transported fragments.

The Central European Sea became progressively more restricted as time
passed (fig. 4.14). This is indicated both by the fauna, which becomes less
diverse up the succession, and by the sediments, because the youngest beds
give evidence of upper subtidal zone deposition interspersed with intertidal phases during which channelling took place. Many elements of the
fauna are also found in the approximately correlative Portland Group
which, as we have seen, also offers evidence of regression up the succession. There are, however, no beds of true Purbeck facies in Bavaria, the
nearest locality where they occur being the Swiss Jura.

Another interesting study relates to the high Upper Jurassic deposits of
northwest Germany (Jordan 1971 *a*; see also Huckriede 1967). Contrary
to previous views of a decrease in salinity starting in the Lower Kimmeridgian, Jordan inferred the reverse. The Gigas–Schichten (Middle
Kimmeridgian) and Eimbeckhäuser Plattenkalk (Upper Kimmeridgian)
are considered transitional in salinity characteristics between the under-

lying fully marine beds and the succeeding Münder Mergel (Upper Kimmeridgian to Lower Cretaceous (?)), which contains halite deposits, because of the presence of anhydrite. An upward decrease in faunal diversity supports the inferred change to conditions of abnormal salinity. The environmental changes outlined by Jordan show an obvious correspondence to those now inferred for the topmost Jurassic deposits in southern England, for which, as described earlier, a comparable change in interpretation has taken place in recent years.

5 Calcareous and siliceous facies in southern Europe and North Africa

In marked contrast to the deposits further north, the Jurassic of the circum-Mediterranean belt is almost entirely calcareous, with limestones predominating over marls, although in the younger deposits radiolarian cherts are an important facies in certain areas. Terrigenous clastic rocks are important only in marginal areas such as western Morocco and parts of the Alpine fold belt in the Balkans. In the last few years our knowledge of the geology has been vastly increased and a fascinating story of palaeogeographic evolution has emerged, involving the break-up and differential subsidence of an extremely shallow water carbonate platform. We shall deal first of all with these platform deposits, which characterise the early part of the Jurassic in most areas, and then go on to consider the deeper water beds which were laid down subsequently. Finally, a palaeogeographic synthesis will be attempted.

CARBONATE PLATFORM DEPOSITS

At the base of the Jurassic in the circum-Mediterranean region, from southern Spain and Morocco to the southern Alps, the Northern Calcareous Alps of Austria, the Apennines and Greece, a thick sequence of massive or bedded limestones and dolomites is developed, usually ranging down into the Triassic without a notable break, and up into the Pliensbachian. The thickness is always substantial and can exceed 1000 m, as locally in the Pantokrator Limestone of Western Greece or the stratigrahically equivalent Calcare Massiccio formation in the Apennines. The deposits include bird's eye, oolitic and pelletal limestones, algal limestones including stromatolitic and oncolitic beds and beds with codiacean and dasycladacean algae, together with massive fine-grained unfossiliferous dolomites and limestones which may contain abundant megalodontid and other types of bivalve, corals or crinoids (e.g. Bernoulli and Renz 1970, Jenkyns 1970a), Bosellini and Broglio-Loriga 1971). The varied facies testify to environments ranging from supratidal to very shallow subtidal. Some of these deposits call for a fuller description.

The modern interpretation of the thin-bedded alternations of laminated unfossiliferous dolomites or limestones with shell-bearing limestones, which are widespread in the early Mesozoic of the Mediterranean region, is due to Fischer (1964). Fischer undertook a perceptive and thorough sedimentological analysis of the late Triassic Dachstein Formation in the Salzburg region and was able to demonstrate convincingly the former existence of algal mats and shrinkage cracks in the laminated beds termed loferites, which were in consequence interpreted as intertidal deposits. The frequent oscillations of these with shallow subtidal shelly limestones were attributed to minor eustatic changes of sea level.

Shinn (1968) has subsequently shown that shrinkage pores that may later become filled by drusy calcite and hence produce a bird's eye texture are more characteristic, in Recent deposits, of the supratidal than the intertidal zone and are never found in subtidal deposits. What Fischer termed Lofer-type cyclothems have been recognised widely in basal Jurassic, besides Triassic, carbonates of the Apennines and elsewhere (e.g. d'Argenio and Vallario 1967, Bernoulli and Renz 1970). Bernoulli and Wagner (1971) have also recognised fossil caliche horizons in deposits of this type in the central Apennines, which consists of pisolitic concretions associated with laminated crusts exhibiting inverse grading. This evidently signifies formation in the vadose zone, and care must be taken to distinguish these subaerially-formed inorganic structures from algal pisolites.

The fullest description of oolitic limestones is given by Fabricius (1967), for the Northern Calcareous Alps near Salzburg. Rhaeto-Liassic oolite bodies there extend in thickness up to 180 m. The Liassic oolites are restricted to a narrow belt, perhaps less than 1 km wide, which is nevertheless as long as 120 km. Lenticular bodies of this sort pass laterally into crinoidal and sponge-bearing limestones, and can be compared to oolite shoals in the Bahamas. Presumably, therefore, the depth of deposition never exceeded a few metres.

Two environmental syntheses involving a wide variety of carbonate platform deposits are worthy of review.

Bosellini and Broglio-Loriga (1971) have undertaken a facies analysis of the Lower and Middle Lias Calcari Grigi in the Rotzo region of the Venetian Alps, which is underlain by late Triassic cyclic lagoonal and tidal flat deposits of the Dolomia Principale.

The following five facies were distinguished, which are more readily recognisable in vertical sequence than laterally:

1. *Oolitic calcarenites*, mainly near the base of the succession, inter-

preted as signifying a barrier island complex comprising subtidal and intertidal bars and shoals; local emergence resulted in the formation of dunes.

2. *Micritic limestones*, rich in molluscs, brachiopods, foraminiferans and algae, deposited in an open lagoon with normal circulation and salinity because of the presence of tidal inlets interrupting the bars of facies 1.

3. *Micritic pelletal limestones*, very poor in fossils, laid down in the most internal part of the lagoon, bordering a series of marshes.

4. *Lithiotis limestones*, comprising lumachelles of this curiously shaped bivalve, which are compared with oyster and mussel banks of Recent lagoons. The vertical or chaotic disposition of the *Lithiotis* shells is held to support this comparison.

5. *Dark grey marls* containing plant remains or even coal seams, signifying a marsh environment.

The association of coals with a marine limestone sequence is certainly unusual; calcareous intertidal and supratidal deposits are not considered to occur in this region.

A classification and interpretation of cavities within the lower part of the Calcari Grigi has been recently attempted by Castellarin and Sartori (1973). They distinguished between smooth-rim vugs due chiefly to penecontemporary desiccation shrinkage processes, and complex branching and minutely irregular-rim vugs due to desiccation shrinkage followed by leaching, or solely to leaching. This work suggests that, contrary to the views of Bosellini and Broglio-Loriga, there are some horizons within the Calcari Grigi which indicate temporary supratidal conditions.

True reefoid deposits are not particularly common in the early Jurassic but Fabricius (1966) recorded the upward continuation of coralgal reef facies from the late Triassic into the Lower Hettangian of the Bavarian and Tyrolean Alps.

A fascinating story has been worked out for the Pliensbachian reef of Jbel bou Dahar, in the eastern High Atlas of Morocco, by du Dresnay (in Agard and du Dresnay 1965). This locality is celebrated for the rich fauna of Tethyan gastropods and bivalves which were described, together with other groups, by Dubar (1948).

Pliensbachian limestones in the High Atlas are usually more or less massively bedded but non-reefoid, although lumachelles of *Lithiotis* are common, as in the Calcari Grigi. Jbel bou Dahar has at its core a massif of Palaeozoic quartzites and schists, known as Sebbab-Kebir, which in Lower and Middle Liassic times formed a topographic high on the sea floor. It was upon this massif that an annular reef was built up, with a lagoon to the south

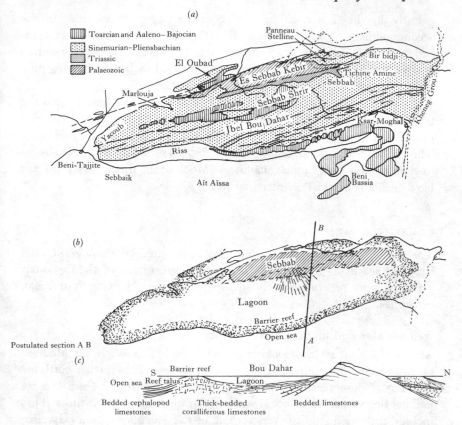

(a)

▥ Toarcian and Aaleno–Bajocian
▦ Sinemurian–Pliensbachian
▩ Triassic
▨ Palaeozoic

Panneau
Stelline
Bir bidji
El Oubad
Es Sebbab Kebir
Tichine Amine
Marlouja
Sebbab
Sebbab Shrir
Ksar-Moghal
Jbel Bou Dahar
Yacoub
Riss
Beni-Tajjite
Beni
Bassia
Sebbaik
Aït Aïssa
Kheneg Grou

(b)

B
Sebbab
Lagoon
Barrier reef
Open sea
A
Postulated section A B

(c)
S   Barrier reef   Bou Dahar   N
Open sea Reef talus   Lagoon
Bedded cephalopod   Thick-bedded   Bedded limestones
limestones   coralliferous limestones

FIGURE 5.1   Interpretation of the Jbel bou Dahar reef complex in the Moroccan
High Atlas. (a) Simplified map. (b) Reconstruction of the reef in the Late
Pliensbachian. (c) Postulated section *A–B*. (Adapted from Agard and du Dresnay
1965.)

incompletely separated from the open sea by a barrier reef (fig. 5.1). The
reef deposits are massive, thickly-bedded limestones containing abundant
colonial corals, some of them *in situ*, and thick-shelled bivalves such as
Opisoma, Daharina and *Gervilleioperna*. The continuity of the reef barrier
is broken by a series of small canyons which du Dresnay interpreted as
contemporary surge channels, rather than modern excavations, because he
had discovered southward-dipping talus deposits flanking the southern
margin of the barrier. Thus the present-day whaleback mountain of Jbel
bou Dahar exhibits an exhumed Liassic relief. To the south of the
mountain the reef deposits can be seen to pass laterally into thin-bedded
limestones, containing little macrofauna but ammonites, which were
evidently deposited in deeper water.

95

Calcareous and siliceous facies in S. Europe and N. Africa

Early in the Toarcian the reef complex began to subside, to the accompaniment of some tectonic fissuring, so that deeper water thin-bedded ammonitiferous Toarcian and Aalenian limestones are now seen to lap over the Pliensbachian deposits. At their base brachiopod-bearing pink Toarcian limestones occupy Neptunean dykes.

A personal examination of the Jbel bou Dahar rocks has revealed the presence of bird's eye and stomatolitic limestone, often pelletal and with foraminiferan, algal and other shell debris, in the interior lagoonal zone. This indicates that du Dresnay's model (fig. 5.1) should be slightly altered to show the lagoon as much shallower. Much of the time, in fact, it must have been a supratidal or intertidal flat.

PELAGIC LIMESTONES

The wide variety of calcareous sediments deposited on a topographically irregular sea floor produced by differential subsidence of the carbonate platform, can be conveniently grouped into the following four major categories.

Red nodular and manganiferous limestones

Among the most striking-looking of the Jurassic deposits in the Mediterranean region, unlike anything in the contemporary beds further north but showing a close resemblance to the Palaeozoic 'griottes', are the red nodular limestones which occur widely in stratigraphically thin units. They include the much-studied Adneterkalk of the Salzburg region (Garrison and Fischer 1969, Hallam 1967b, Hudson and Jenkyns 1969, Jurgan 1969, Wendt 1971b) and the celebrated Ammonitico Rosso of the Southern Alps and Apennines. They also occur in North Africa, the Subbetic zone of southern Spain (Geyer 1967a), the Carpathians (Misik 1964, Szulczewski 1965), the Bakony Mountains of Hungary (Geczy 1961) and western Greece (Bernoulli and Renz 1970).

Besides serving as a highly attractive ornamental stone these limestones pose intriguing problems of interpretation and have engendered much discussion, especially on the red coloration, the nodularity and depth of deposition.

As regards the composition, the clay fraction is almost pure illite and the red coloration due to a small percentage of haematite. The iron content is no higher than in the correlative bluish-grey smooth-bedded limestones that will be described in the next section, but these latter contain small quantities of pyrite and organic matter which are absent from the nodular limestones. It is evident that the red coloration is a secondary, diagenetic phenomenon and that iron oxide, probably in the form of goethite, was precipitated in a strongly oxidising environment deficient in organic matter, and was subsequently dehydrated to haematite. There is no need to invoke, as has been done on a number of occasions, a lateritic source to account for the colour. Actually, the deepest reds are restricted to the marlier rock, the limestones being more characteristically pink or even, if almost pure, pale grey. In other words the proportion of $CaCO_3$ exercises a major influence on the coloration (Hallam 1967 *b*).

The origin of the calcareous part of the rock is more uncertain. The texture ranges from micritic to microsparitic, so that electron microscopy is required to tackle the problem. This reveals the presence of nano-organisms including coccoliths (Garrison and Fischer 1969, Bernoulli and Renz 1970), which become important from the Toarcian onwards, and the incertae sedis *Schizosphaerella* (Bernoulli and Jenkyns 1970, Jenkyns 1971 *b*). Much of the material, however, consists of inorganic calcite which has crystallised interstitially, so that it has proved impossible, as yet, to assess the relative importance of the biogenic fraction. According to Jenkyns (1971 *b*) the rarity of obvious nanofossils is probably due mainly to mechanical and solutional influences effecting their disintegration, but in addition it is possible that prior to the late Jurassic many nano-organisms might have not been fully calcified.

As the name Ammonitico Rosso suggests, ammonites in the form of moulds are an important component of these deposits and may be extremely abundant. Valves of the presumed pseudoplanktonic bivalve *Bositra* are common from the Toarcian to the Oxfordian but other macrofauna is generally rare. The sparse benthonic element includes pygopid brachiopods, crinoid ossides and fish teeth, though small gastropods and other invertebrates may be abundant in highly condensed fissure deposits, as will be described below. Trace fossils are restricted to the abundant *Zoophycos* and *Chondrites*. The microfauna is dominated by calcitised radiolaria and also includes calcitised sponge spicules, globochaetes and rarer foraminiferans and ostracods.

A useful distinction has been made by Aubouin (1964) between the stratigraphically thicker and more marly *Ammonitico Rosso marneux*, considered to be deposited on the flanks of submarine swells, and the *Ammonitico Rosso calcaire*, deposited on the tops of swells. The former is gradational to the basinal sequences to be described in a later section and may contain allodapic horizons of shallow water skeletal material such as crinoid debris or resedimented pelagic *Bositra* (e.g. Bernoulli and Jenkyns 1970). Condensation of the latter, where the argillaceous content diminishes to a few per cent, can be extreme, with several Jurassic stages being reduced in thickness to a few metres or less, as for instance in western Sicily or in the Trento–Lake Garda region of the southern Alps. Such highly condensed beds exhibit a number of interesting characteristics in addition to those already described (though it should be noted that nodularity is not so marked as in the less condensed beds).

Manganese in the form of black pyrolusite is often concentrated, in association with brown limonite, either as laminated crusts of varying thickness on hardgrounds or ammonite moulds, or as concentrically laminated nodules. Jenkyns (1970c) has undertaken a detailed study of such nodules in the Middle Jurassic of Sicily and convincingly demonstrated their close resemblance in shape, internal structure, mineralogy and geochemistry to manganese nodules occurring on the present-day oceanic seamounts. Strictly speaking the Sicilian occurrences are *ferromanganese* nodules, because the iron content is high. Jenkyns tentatively relates this to hydrothermal effusions because of an association with volcanic rocks. At least some of the manganese, however, has probably been derived from continental areas. In ferromanganese crusts and nodules in the Tyrol, in which the whole of the Jurassic from the Sinemurian to the Oxfordian inclusive is reduced to less than 17 m, Wendt (1969) has recognised encrusting sessile foraminiferans and serpulids and the borings of algae. Occurrences of limonitic crusts capping hardgrounds and associated with stratigraphic breaks in the Spanish Subbetic zone are described by Geyer and Hinkelbein (1971).

Further evidence of algal activity comes from stromatolites, which have been recognised both in western Sicily and the Trento region, and which consist characteristically of flat laminae or rounded, flattened domes of the LLH type of Logan *et al.* (1964); see Sturani (1971) and Jenkyns (1971b). Discussion on the palaeogeographic significance of these structures and the ferromanganese concentrations will be postponed until later in the chapter.

Among the most remarkable deposits of the Mediterranean Jurassic are the submarine fissure fillings described from the Austrian Alps and western Sicily by Wendt (1971a). An especially fascinating story emerges from Wendt's detailed and penetrating analysis of the fissures of Rocca Busambra in the latter region, which range in age from Toarcian to Miocene and occur in massive Liassic limestones of the carbonate platform facies.

Wendt distinguished Q fissures, which are oblique to the bedding and may descend hundreds of metres, from S fissures which are parallel to the bedding. Certain of the S fissures have filled extremely slowly. As a result of repeated opening along the former roof of the cavities, the Rocca Busambra fissures were able to accommodate a sediment sequence ranging

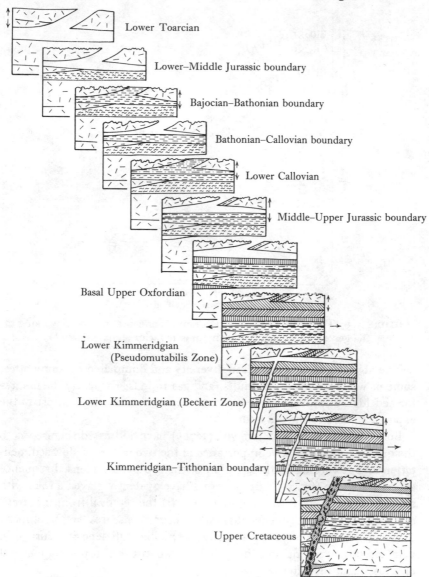

Lower Toarcian

Lower–Middle Jurassic boundary

Bajocian–Bathonian boundary

Bathonian–Callovian boundary

Lower Callovian

Middle–Upper Jurassic boundary

Basal Upper Oxfordian

Lower Kimmeridgian
(Pseudomutabilis Zone)

Lower Kimmeridgian (Beckeri Zone)

Kimmeridgian–Tithonian boundary

Upper Cretaceous

FIGURE 5.2 Schematic representation of phases of evolution of submarine fissures in the Jurassic of Rocca Busambra, Sicily. (Adapted from Wendt 1971 *a*.)

in age, as proved by ammonites, from Lower Toarcian to Lower Kim-meridgian and which is only one metre thick (fig. 5.2). This is actually the most complete ammonite succession known in the Mediterranean Jurassic!

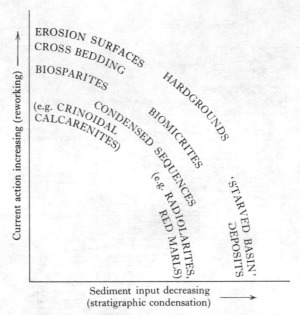

FIGURE 5.3 The relationship of stratigraphic condensation and reworking to the genesis of various Tethyan Jurassic facies (Jenkyns 1971 *b*).

The abundant fauna is high in diversity and dominated by ammonites, some of which are dwarfed adults, and gastropods. Evidently the fissures formed highly favoured sites for fossil accumulations, in contrast to the exposed surface of the sea floor.

Both Wendt (1970) and Jenkyns (1971 *b*) have addressed themselves to the question of the relative importance of the two factors of slow sedimentation and reworking in the creation of the types of condensed sequence described above (fig. 5.3). Both factors have evidently played a role, with specific evidence for reworking including (i) the removal from the stratigraphic record of sediments that only occur in fissures or 'Neptunean dykes', (ii) scattered eroded fragments of a Sicilian tuff deposit occurring in beds directly overlying and (iii) differences between the infilling material of a fossil and the external matrix.

One of the thorniest problems of all for this general category of limestones is the origin of the nodules. Two clearly distinct alternative views have been put forward. The first, which attributes the nodule formation to some form of submarine solution, was proposed by Hollmann (1964) in his account of the Upper Jurassic Ammonitico Rosso Superiore of the Trento region, and has been widely accepted (e.g. Garrison and Fischer 1969, Jurgan 1969).

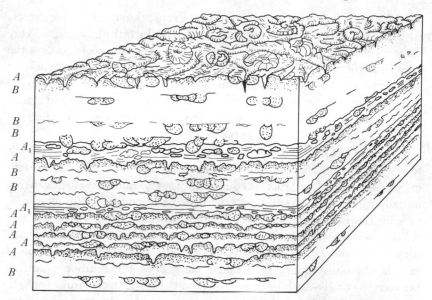

FIGURE 5.4 Corrosion surfaces in the Ammonitico Rosso Superiore in northern Italy. *A*, hardgrounds; *B*, corrosion surfaces in only partly consolidated sediment; A_1, corroded limestone nodules in marls. Ammonites are stippled (Hollmann 1964).

According to Hollmann, the phenomenon of 'subsolution', on or just below the sea bed, accounts not only for hardgrounds and the corroded upper surfaces of ammonites (fig. 5.4) but for the nodules as well, which are envisaged as the undissolved remnants of limestone beds, left suspended in an insoluble residue of marl. It is very difficult to conceive, however, how such extensive solution, of calcite as well as aragonite, could have taken place in deposits which contain stromatolites. The fact that skeletal concentrations are greater in the marly interstices than the nodules, which has been cited as favouring the differential solution hypothesis, is indecisive, because it can just as readily be explained by differential compaction. Likewise, the existence of stylolitic margins around some of the nodules indicates local small-scale solutional effects but does not prove anything about the origin of the nodules themselves.

The alternative view is that the nodules formed by early diagenetic segregation (Hallam 1967 b, Hudson and Jenkyns 1969). Significant in this respect in the fact that nodular limestones are characteristic of stratigraphically condensed sequences, suggesting an association with slow sedimentation rates. In the light of modern discoveries in the Persian Gulf and elsewhere, this might indicate that diagenetic crystallisation around certain nuclei will take place spontaneously if the pace of sedimentation is kept at a low level. Evidence from the Scheck, a redeposited submarine breccia with sparry calcite cement occurring in the Adneterkalk, indicates very early formation of the nodules (Hudson and Jenkyns 1969). So does the occurrence of plastically deformed or 'streaked out' nodules in redeposited sediments of the marly Ammonitico Rosso facies. Other facts to be accounted for include the gradational contacts between nodules and matrix and the preservation of originally siliceous sponge spicules and radiolaria, whether calcite-replaced or not, only in the nodules.

Jenkyns (1974) has recently put forward a comprehensive hypothesis to account for the formation of nodules in regular bands alternating with marlier sediment, by a process of diagenetic segregation. It is envisaged that initially much aragonite was present which was subsequently dissolved, causing the interstitial waters to become supersaturated with respect to calcite, which was precipitated at sites probably determined by the presence of skeletal calcite grains. In addition, some selective solution of calcitic coccoliths might have taken place once the solution gradients between the newly-established nodules and the matrix were strong enough. At a certain depth below the sediment–water interface bacteria might have removed protective organic coatings from skeletal particles, hence facilitating carbonate solution.

Compared with shell fragments or micritic intraclasts, the nodules were able to compete favourably for calcite. Initially, when the sites of nodule growth were close to the supply of carbonate, rapid diffusion to a variety of sites must have ensured that the critical supersaturation level for skeletal calcite was not reached. As the sediment–water interface migrated upward, the concentration gradient and hence the rate of downward ionic diffusion would have decreased. At a certain point the critical level for skeletal calcite was reached. A new set of nodules began to form higher up in the sediment as the rate of withdrawal of $CaCO_3$ by the earlier-formed nodules decreased but the supply of soluble carbonate near the sediment–water interface continued. By this means a form of post-depositional rhythmic precipitation could have taken place, which has an obvious bearing on the origin of the limestone–marl rhythms discussed in the previous chapter. Whether Jenkyns' hypothesis or some development of it will prove generally applicable is uncertain, especially as probably 'secondary' limestone–marl sequences exist which contain very little skeletal calcite. However, the preservation of skeletal calcite may not be critical for the precipitation process.

The solution of siliceous microfossils in the matrix implies undersaturation with respect to silica in the ascending pore waters. In zones where nodules were formed these microfossils were protected.

Skeletal limestones

In western Sicily, at the top of the carbonate platform sequence and directly below condensed pelagic limestones of deeper water facies is a horizon of Middle Liassic crinoidal limestone, discussed in detail by Jenkyns (1971 a). Jenkyns inferred the development of a series of crinoidal carbonate sand waves on the tops of current-swept seamounts and drew a parallel with globigerinid and pteropod sand bodies which are found today on the Blake Plateau. The presence of kaolinite was thought probably to signify the exposure of lateritised areas on oceanic islands. This interpretation might well have more general significance for the Mediterranean region because Jenkyns considers that these crinoidal limestones are characteristic of the transition from carbonate platform to pelagic facies. Some of these have karst-like features.

Lumachellic deposits of Bajocian age mark this transition in the Venetian Alps, for instance, but in this case they are enriched primarily by shells of *Bositra* and other molluscs and hence are known as the *Posidonia alpina* beds. These have been analysed in detail by Sturani (1971).

The deposits were interpreted as a coquina which became partly trapped in fissures in the underlying massive Liassic limestones, probably having been swept there by hurricane-generated storm waves. The small size of the fossils was attributed partly to mechanical sorting and partly to biological factors. An algal meadow environment on a rocky sea bed was envisaged, as signified by the abundance of minute herbivorous gastropods (modern algal meadows often have diminutive epifaunal invertebrate species living in association). As regards the ammonites, only juveniles of phylloceratids and lytoceratids are evidently present, suggesting that the algal meadows might have served as spawning grounds for these groups. Other ammonite groups are represented, however, by mature individuals. As in the case of the Sicilian Jurassic, these deposits are directly overlain by highly condensed pelagic limestones, indicating an environment of minimal sedimentation.

Brief mention should also be made of the other well-known crinoidal limestones, such as the Lower Liassic Broccatello of the Southern Alps and the Hierlatzkalk of the Northern Calcareous Alps.

Flat-bedded grey marly limestones with chert concretions

These are volumetrically by far the most important rocks in the pelagic facies of the Mediterranean Jurassic. In the region of the southern Alps between Lakes Como and Lugano, for instance, the Lower Lias reaches almost 4000 m (Bernoulli 1964), while in the Allgau Beds in Austria the

Upper Pliensbachian alone is estimated at nearly 1000 m (Jacobshagen 1965). The rocks are regularly-bedded micrites or microsparites with marl partings and scattered bands of secondary chert nodules. There are more or less frequent intercalations of limestone turbidites (or allodapic limestones) containing skeletal debris from a shallow water environment. These, together with less frequent sedimentary breccias and slump folds, attest to submarine slopes of appreciable gradient. The sparsity or almost total absence of benthos supports the inference from these sedimentary features of basinal deposition in water of at least moderate depth.

Bernoulli (1971) has recognised a special category of beds, which he described as redeposited pelagic sediments, whose components, such as valves of *Bositra*, ammonites, rare benthonic foraminiferans and ostracods and occasional echinoderm remains, do not differ much from those of the enclosed formations. Therefore they indicate an intrabasinal source of supply and the activity of basinal bottom currents, which are also indicated by cross laminations and laminations parallel to the bedding.

Two groups of rocks sharing the features noted above can be distinguished. In rocks older than the Tithonian, dark to light grey argillaceous limestones with marly intercalations often exhibit a mottling due to trace fossils including *Chondrites* and *Zoophycos*, hence the names *Fleckenkalk* and *Fleckenmergel* used in German-speaking countries. Such beds usually reveal in thin section an abundance of radiolarian and sponge spicules (hence *Spongienkalk*) which may be calcitised or recrystallised as chalcedonic quartz. The more chert-rich varieties have been termed *Hornsteinkalk*.

One such sequence of Lower Liassic age in the Glasenbach Gorge a few kilometres south of Salzburg was examined in detail by Bernoulli and Jenkyns (1970). Electron microscopy revealed the presence of nano-organisms of uncertain affinities (e.g. *Schizosphaerella*), but the bulk of the limestone showed no obvious organic structure. Coccoliths only became important from the Toarcian onwards.

The second category consists characteristically of near-white or light grey lithographic limestones, of Tithonian age and ranging without a break up into the Cretaceous. This includes the celebrated Biancone and Maiolica of the southern Alps and Apennines, the Vigla Limestone of Greece, and the Oberalm Beds of Austria. The macrofauna is virtually confined to ammonite aptychi and rare *Pygope* but there is a much richer microfauna, including calpionellids, plates of the pelagic crinoid *Saccocoma* and the presumed planktonic alga *Globochaete*. Electron microscopy reveals that coccoliths are a major component of these deposits (Farinacci 1964, Flügel 1967, Garrison 1967).

For more detailed information we shall consider the thorough study by Garrison (1967) on the Oberalm Beds of the Unken syncline in Salzburg Province. Garrison distinguished four types of deposit:

1. *Pelagic limestones.* These occur as beds ranging in thickness from 2 to 50 cm, with an average of 10 cm, and contain secondary chert nodules. According to Flügel and Fenninger (1966) the $CaCO_3$ content ranges mostly from 85 to 90 per cent. The microfauna and flora is as described above; the micrite which comprises the bulk of the rock is composed of coccoliths and recrystallised calcite. Careful observation revealed the presence of redeposited pelagic limestones testifying to the action of bottom currents. These are laminated and may show grading of radiolarian tests. Their soles may be plastered with reworked aptychi.

2. *Marls.* These are thought to signify either episodes of reduced plankton productivity or the fine-grained tops of turbidites. Garrison failed, however, to consider the possibility that they may owe their origin to rhythmic segregation in diagenesis.

3. *Allodapic limestones* (term of Meischner 1964). Grading is present in many of these turbiditic beds, which range up to 2 m thick and contain a variety of shallow water material including fragments of dasycladacean algae, crinoids and molluscs, together with ooids and rarer quartz grains. Flute casts on the soles signify currents from the northeast. Secondary chert occurs mainly in this type of rock, presumably because the higher porosity allowed more ready access of silica-bearing fluids and more interstitial space for precipitation. The source of silica is to be found in the solution of radiolarian tests and sponge spicules.

4. *Limestone breccias.* These are but a minor component of the Oberalm Beds; they exhibit some graded bedding and contain angular allochthonous fragments thought to be derived, like the other allodapic material, from submarine rises to the north and south, which shed sediment into an east–west oriented trough.

Pelagic 'oolites' and iron-rich pisolites

These are significantly different types of rock but the ironstones are not sufficiently important to warrant a separate category.

In the Upper Jurassic of Sicily are certain beds which contain structures resembling ooliths, but electron microscopy reveals that they have cortices of calcitic nano-organisms. Hence the micritic layers must have grown, at least partly, by particle accretion, in contrast to Bahamian ooliths which grew by the precipitation of aragonite crystals. Jenkyns (1972) therefore interpreted these Sicilian 'ooliths' as micro-oncolites produced by the activity of blue–green algae, an interpretation which is supported by the association with stromatolites.

Unlike in many shallow water Jurassic oolites, such as those in the Bajocian of the English Cotswolds, there is no secondary sparry calcite that has grown radially during the recrystallisation of the originally aragonitic ooliths. The Sicilian deposits, containing as they do globigerines and *Saccocoma*, evidently signify more open sea and deeper water conditions than on the Great Bahama Bank where planktonic micro-organisms are extremely scarce. Similar rocks occur in the Jurassic of Hungary and Poland and have also been interpreted as micro-oncolitic deposits (Radwanski and Szulczewski 1966).

Another interesting Sicilian occurrence recognised by Jenkyns (1970*b*) is an iron-rich pisolite of Toarcian age, which is interbedded with condensed pelagic red limestones. This contains limonite, haematite and rare chamosite, the limonite infilling, in some cases, algal or fungal bores. Because the pisolite contains clasts of trachyte it is probable that much of the iron and trace elements were derived from the hydrothermal effusions that accompany submarine extrusions.

RADIOLARITES

In the Upper Jurassic of many areas pelagic limestones pass up into grey–green or reddish-brown cherts, of no great stratigraphical thickness, containing the remains of abundant Radiolaria. These so-called radiolarian cherts or radiolarites were well described by Grunau (1965) and Garrison and Fischer (1969).

The cherts are characteristically thin and smooth-bedded, the beds being separated by wafer-thin partings of shale. The poorly-preserved radiolarian tests, which can comprise up to 70 per cent of the rock, are filled with chalcedony or secondary calcite, most of the original test walls having been dissolved during diagenesis. Sponge spicules may be quite common, and aptychi, crinoid ossicles and *Bòsitra* valves occur rarely; otherwise the beds are barren of fossils. Synsedimentary slumping is sometimes recognisable.

Although these rocks lack interesting sedimentary structures or textures and are virtually devoid of macrofauna they have attracted much attention in the past, in particular because of the well-known association with ophiolites. Such an association is evidently not the case for deposits such as the Ruhpolding radiolarite of Austria, which occurs within a pelagic limestone sequence totally lacking in igneous rocks, and many similar formations in the southern Alps and elsewhere. Discussion of their considerable palaeogeographic significance is best dealt with in the next part of the chapter.

Occasionally, turbiditic clastic beds are found associated with the Upper Jurassic radiolarites, as in parts of the Northern Calcareous Alps of Austria (Schlager and Schlager 1973). In the example in question, the clastic material consists of marls and limestones of Rhaetian and Jurassic age, probably derived from a tectonically uplifted zone. The clastics are thought to have been deposited on submarine fans bordering this zone. Slides, slumps, mudflows and fluxoturbidites have been recognised, and the general facies resembles that of many flysch troughs.

PALAEOGEOGRAPHIC EVOLUTION

The question of the depth of deposition of the pelagic rocks has proved a vexed one, with controversy continuing right up to the present day. The presence of stratigraphic gaps, supposed desiccation features and scattered dolomite crystals has been held to signify shallow water deposition with periodic emergence for condensed limestone sequences of the central Apennines (Colacicchi and Pialli 1967, 1971, Farinacci 1967). The deposits in question, however, show none of the features that are now universally accepted as unequivocal indicators of deposition in very shallow water, or emergence into the intertidal or supratidal zone. Moreover, dolomite has been discovered in Recent deep sea environments (Fischer and Garrison 1967). By general agreement the pelagic limestones, marls and cherts were laid down in significantly deeper water than the deposits of the underlying carbonate platform. Just how much deeper is a question of considerable sedimentological and geotectonic significance.

Many workers have favoured the attainment of oceanic depths of up to several thousand metres by the early part of the late Jurassic (e.g. Merla 1952, Trümpy 1960, Geczy 1961, Misik and Rakus 1964, Colom 1967). Much the most thorough and lucid analysis, utilising modern knowledge, which draws this conclusion is that of Garrison and Fischer (1969), discussing the Jurassic of the Salzburg region; their work must accordingly be considered at some length.

Garrison and Fischer used four lines of argument for bathyal to abyssal depths:

(i) *The fossil content.* The sparsity of benthos and the abundance of planktonic microfossils are consistent with the sharp diminution of suspended food with increasing depth of the water column observed in the present day ocean system. This results in biological starvation on the deep ocean floor, with a consequent reduction in benthonic faunal density.

(ii) *Slow sedimentation rates.* The considerable or even extreme condensation of some of the pelagic limestone sequences has already been described. This is thought to be consistent with sediment starvation in deep water, corresponding to Trümpy's (1960) *leptogeosynclinal phase* of Alpine evolution. Obviously, below the photic zone the productivity of calcareous algae will diminish to zero and that of invertebrates with calcareous skeletons diminish with depth appreciably. Inorganic precipitation in cooler, deeper water, undersaturated with respect to $CaCO_3$, will have been out of the question and, prior to the nanoplankton evolutionary burst at the close of the Jurassic, the contribution of coccoliths to deep water sediments would have been limited.

(iii) $CaCO_3$ *solution.* Hollmann's interpretation that the nodular limestones have undergone considerable solution is accepted, and held to signify deposition in water considerably undersaturated in calcium and biocarbonate ions, i.e. deep water. The corrosion of ammonites and lack of calcitic replacements of their originally aragonitic shells is contrasted with ammonite preservation in shallow water deposits. The lack of ammonite moulds in the Oberalm Beds and their stratigraphic equivalents, where ammonite shells must have settled because of the evidence of aptychi, indicates aragonite solution on the sea bed before sufficient time had elapsed for the shells to be filled by sediment.

With regard to the radiolarites, these have long been held to be deep oceanic deposits. Two alternative hypotheses have been put forward to account for the silica enrichment. One relates this to volcanic exhalations on the sea bed, but, as noted earlier in this chapter, the radiolarites in question occur in a normal stratigraphic sequence of sedimentary rocks and lack the close association with pillow basalts that the hypothesis requires. The only reasonable alternative, accepted by the authors, is that the radiolarites were laid down below the $CaCO_3$ compensation depth and are thus ancient analogues of radiolarian ooze.

In the present oceans the calcite compensation depth fluctuates between about 4500 and 5500 m, its level depending on a number of variables, principally the balance between the rate of supply and solution of the mineral. Aragonite dissolves in significantly shallower depths than calcite. Garrison and Fischer accepted that the compensation depth of both calcite and aragonite in the Jurassic Tethys might have been somewhat shallower, either because of higher temperatures leading to higher chemical reaction

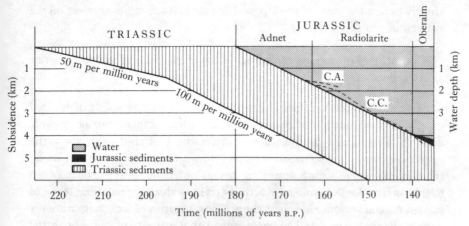

FIGURE 5.5 Interpretive model of the bathymetric evolution of the southern and eastern Alps during the Triassic and Jurassic. C.A and C.C. are aragonite and calcite compensation depths (Garrison and Fischer 1969).

rates (cf. Hudson 1967) or because the rate of supply of calcareous micro-and nanoplankton to the deep sea floor was appreciably less than today. Nevertheless, the depths of deposition proposed are considerable. For the Adnet Limestone a depth of around 4000 m was suggested, for the overlying Rudpolding radiolarite 4500–5500 m, and for the Oberalm Beds 4100–4500 m.

(iv) *Subsidence rates.* The underlying late Triassic and early Jurassic carbonate platform deposits are several thousand metres thick over wide areas. Since they were evidently all deposited close to sea level, substantial rates of subsidence, of the order of 100 m per 10^6 years, can be inferred. Extrapolation upwards into the younger Jurassic gives an estimate of the depth of the marine basin created, on the assumption that the subsidence rate remained constant but that the rate of sediment production diminished dramatically (to less than 1 m per 10^6 years). The figure arrived at is very similar to that inferred from the argument based on the compensation depth.

Alternative evolutionary models were proposed. The first is based on the assumption that the compensation depth was slightly shallower than in the oceans today (fig. 5.5) and the second assumes no such difference. This latter model therefore implies a sharper initial rate of subsidence, followed by a slight rise of the sea bed in the Upper Jurassic. It is accepted in both models that the evolutionary burst of calcareous micro- and nanoplankton shortly before the close of the Upper Jurassic had a significant effect on the pattern of sedimentation.

A number of objections can be made to Garrison and Fischer's impressively argued case.

(i) The sparsity of benthos is no doubt primarily the result of limited supply of suspended food particles in the form of plankton or organic detritus. This indeed diminishes with increasing depth but also towards the open ocean away from coastal regions into which nutrients are carried by rivers. Only where strong upwelling brings nutrients from the deep ocean to the surface does organic productivity rise to values approaching those of certain coastal regions. Much of the tropical ocean is in fact comparatively barren (Ryther 1963). It has been argued that sparsity of benthos in the Ammonitico Rosso and Adneterkalk could be a consequence of the increased pelagic influence in waters of only moderate depth, compared with the shallower, more inshore waters of northern Europe (Hallam 1967 b).

(ii) The postulate that deep marine basins may be starved of sediment compared with shallower waters needs considerable qualification. In the present day ocean, net sedimentation rates tend to be lowest on swells or rises and highest in the intervening basins, as currents tend to sweep off material into deeper water. A further point to note is that Moore (1970) has shown that suspended terrigenous clay particles within a bottom layer no more than a few metres thick can give this water layer sufficient density to flow slowly basinward as a low velocity density current. Further, he has shown that these density currents tend to follow bathymetric lows and flow around bathymetric highs. This could account for the fact that the Jurassic basinal sequences are more marly and the swell deposits richer in calcareous pelagic components because the clay would have settled preferentially in the topographic depressions (cf. Aubouin 1964). Garrison and Fischer's hypothesis provides a most implausible explanation of the examples of extremely slow sedimentation rates in Sicily and the Venetian Alps, where several Jurassic stages are represented by only a few metres or even less.

The hypothesis is countered more decisively by the evidence from widely separated regions of algal activity in the condensed red limestone sequences, as provided most positively by stromatolites, but also plausibly inferred from the lumachelles rich in herbivorous gastropods and from boring activity in beds in Austria, as described by Wendt (1969). Wendt argued that the borings signify the activity of algae rather than fungae (with which they can be easily confused) because they do not occur on the upper surface of bedding-parallel S fissures, where illumination would have been totally lacking.

The implication of this evidence is, of course, that deposition must have taken place within the photic zone, which today never descends below about 150 m even in clear waters of low latitudes. It seems likely that those beds with good stromatolites were laid down at depths appreciably shallower than this, but equally probably much of the marlier Ammonitico Rosso was laid down at greater depths. The important point is that depths obtained of the order of a few hundred rather than a few thousand metres.

(iii) If the formation of nodules within the Ammonitico Rosso and equivalent formations is attributed not to the mechanism proposed by Hollmann but to diagenetic segregation, as argued earlier in this chapter, then the case for significant solution of $CaCO_3$ is severely weakened. As for the solution of ammonite and other aragonitic shells either on or within the sea bed, there is evidence from present-day seas that this can take place at depths of only a few hundred metres (Jenkyns 1971 b).

The case of the radiolarites is an intriguing one, because it is difficult to see how such deposits could have formed unless $CaCO_3$ were eliminated by solution, i.e. below the compensation depth. The factors controlling the solution and precipitation of $CaCO_3$ in present-day seas are all too little understood, however, for us to apply our knowledge of its present depth directly to the Jurassic Tethys. As Hudson (1967) has argued, a speeding up of chemical reaction rates in the more equable Mesozoic climate might have had the effect of causing a significant shallowing of the compensation depth. Furthermore, the replacement in Tithonian times of radiolarites by pelagic limestones almost certainly relates to the evolutionary radiation and prolific expansion in numbers of calcite-secreting planktonic foraminiferans and algae, and there need have been no late Jurassic shallowing as postulated in one of Garrison and Fischer's alternative models. This important biological change must have led to a significant depression of the compensation depth.

A further objection to Garrison and Fischer's hypothesis is that oceanic depths of several thousand metres generally imply oceanic crust, yet this seems to be ruled out in the Mediterranean rocks under discussion, which follow conformally on a shallow water carbonate sequence in turn resting upon marine and non-marine clastics with a granitic basement (Trümpy 1960). The alternative implies drastic thinning of the continental crust or

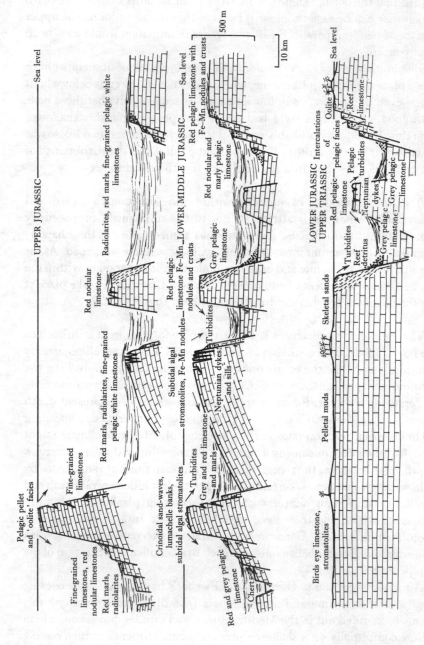

FIGURE 5.6 Stages in the collapse of the Tethyan Jurassic carbonate platform (Bernoulli and Jenkyns 1974).

'oceanisation', which, although it cannot be decisively ruled out, has still to be convincingly demonstrated as a major process. It is true that deep JOIDES drillings in the western North Atlantic have revealed late Jurassic and early Cretaceous sediments, presumably lying on oceanic crust, which bear a close resemblance to unconsolidated Ammonitico Rosso and Maiolica (Bernoulli 1972). All this signifies is that such deposits may be laid down over a considerable depth range, and does not exclude shallower depths for the Jurassic. Moreover, there is no way of knowing at present how much these Atlantic occurrences might have subsided after deposition.

The alternative evolutionary model for the Mediterranean region which is preferred here also involves considerable subsidence and takes note of the evidence of tectonism operating from Liassic times onwards. This evidence takes several forms. Considerable lateral variations in stratigraphic thickness, from condensed red limestones to thick units of grey cherty limestones, often take place over short distances, and slump fold signify appreciable submarine slopes. Sedimentary breccias indicative of contemporary fault activity are widespread, and are especially well documented for the southern Alps (Wiedenmayer 1963, Bernoulli 1964). Neptunean dykes are sediment-filled fissures signifying tension and have been recorded from Morocco (Agard and du Dresnay 1965), Sicily (Wendt 1971a), the Venetian Alps (Castellarin 1965) and the Tyrol (Wendt 1969). Contemporary small-scale volcanism, indicated by intercalations of lava, tuff or bentonite, are likewise recorded from southern Spain (Paquet 1969), Sicily (Jenkyns and Torrens 1971) and the Venetian Alps (Bernoulli and Peters 1970).

Integration of all the different types of evidence, sedimentary, faunal and tectonic, leads to the reconstruction of an extensive shallow water carbonate platform which was established by late Triassic times. During the early and middle Jurassic it underwent tensional stresses and broke up into a series of blocks which subsided differentially through the remainder of the period (fig. 5.6). Certain regions became converted into basins which accumulated fine-grained pelagic calcareous or siliceous sediment. Others resisted subsidence to a less or greater extent, either persisting as smaller platform regions like the Great Bahama Bank or becoming converted into fault-bounded seamounts rising to within a few hundred metres of sea level, with a sedimentary environment comparable in many respects to the Blake Plateau, but smaller. From time to time shallow water carbonate sediments were shed into the basins as turbidites, and periodic activity along the bordering fault scarps yielded debris which descended the slopes to give rise to breccia deposits. More siliceous and organic matter accumulated within

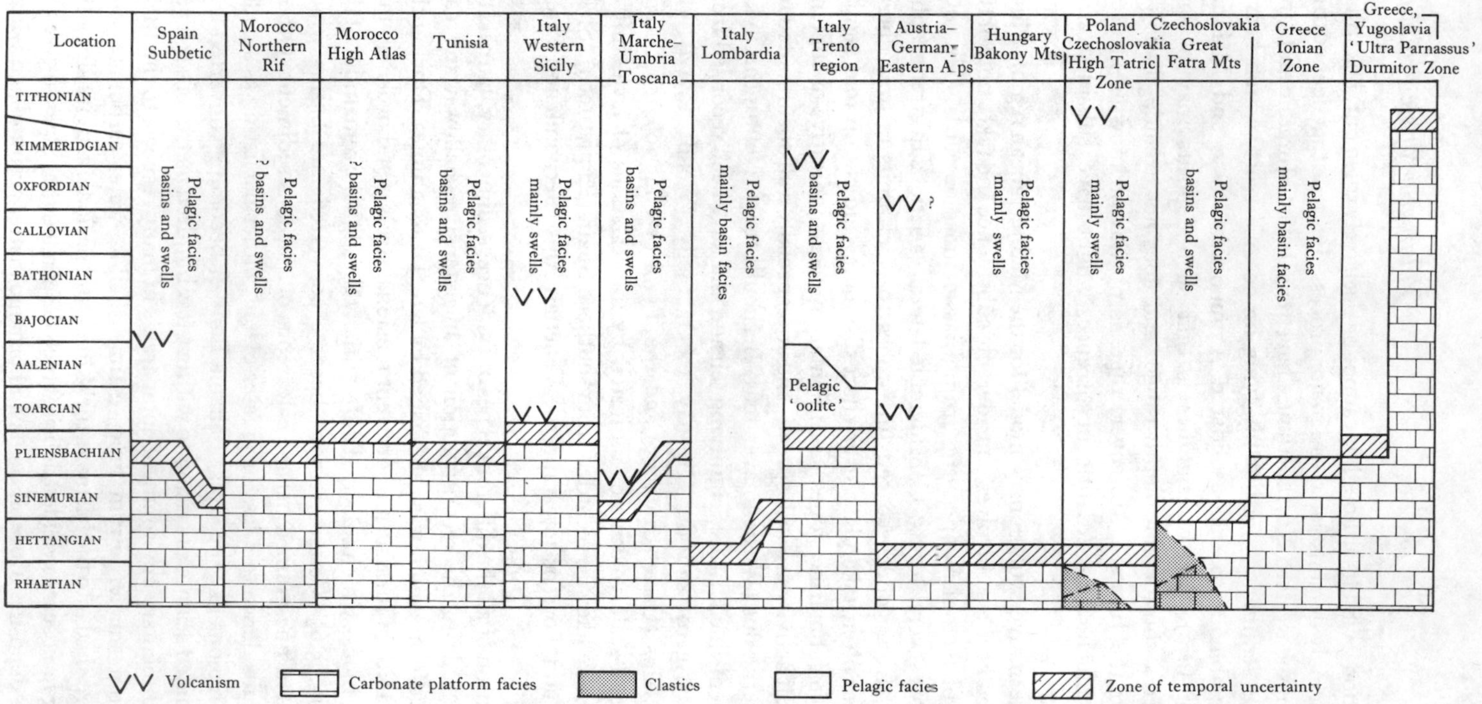

FIGURE 5.7 Facies changes in the Tethyan Jurassic of the peri-Mediterranean region related to the stratigraphic sequence (Bernoulli and Jenkyns 1974).

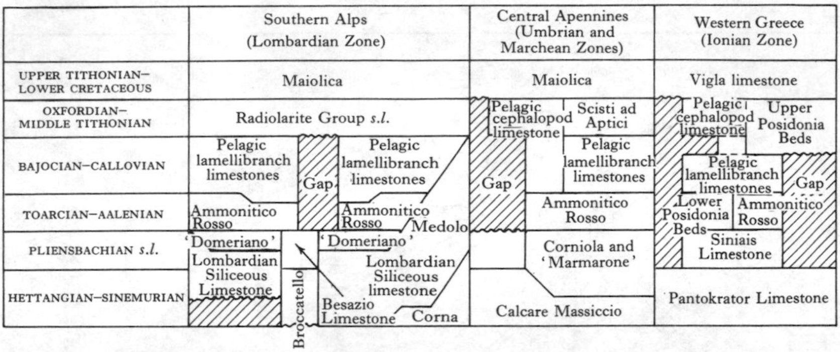

	Southern Alps (Lombardian Zone)			Central Apennines (Umbrian and Marchean Zones)	Western Greece (Ionian Zone)
UPPER TITHONIAN–LOWER CRETACEOUS	Maiolica			Maiolica	Vigla limestone
OXFORDIAN–MIDDLE TITHONIAN	Radiolarite Group *s.l.*			Pelagic cephalopod limestone | Scisti ad Aptici	Pelagic cephalopod limestone | Upper Posidonia Beds
BAJOCIAN–CALLOVIAN	Pelagic lamellibranch limestones	Gap	Pelagic lamellibranch limestones	Gap | Pelagic lamellibranch limestones	Pelagic lamellibranch limestones | Gap
TOARCIAN–AALENIAN	Ammonitico Rosso		Ammonitico Rosso / Medolo	Ammonitico Rosso	Lower Posidonia Beds | Ammonitico Rosso
PLIENSBACHIAN *s.l.*	'Domeriano' Lombardian Siliceous Limestone		'Domeriano' Lombardian Siliceous limestone	Corniola and 'Marmarone'	Siniais Limestone
HETTANGIAN–SINEMURIAN		Broccatello	Besazio Limestone | Corna	Calcare Massiccio	Pantokrator Limestone

FIGURE 5.8 Simplified stratigraphic scheme of the Jurassic sequences of the southern Alps, the central Apennines and western Greece (Bernoulli 1971).

the basins than on the highs as in comparable environments also at the present day (Lisitsyn 1967).

From a review of the stratigraphic literature Bernoulli and Jenkyns (1974) concluded that break-up and differential subsidence in the Mediterranean region extended over a considerable span of time but reached a peak in the Pliensbachian and Toarcian (fig. 5.7).

A clearer picture of the palaeogeographic changes should emerge from a more detailed account of the more thoroughly studied regions.

Western Sicily. The growth and subsequent disintegration of the carbonate platform is well reviewed by Jenkyns (1970*a*) and Jenkyns and Torrens (1971).

Above the Dolomia Principale, of Triassic and basal Jurassic age, come a series of dolomites and more or less massive limestones containing Lofer-type cycles and other carbonate platform facies. Platform deposition ceased in the Pliensbachian, and in a short stratigraphic distance crinoidal biosparites capping platform deposits pass up into Toarcian grey pelsparites, red micrites and iron pisolites. Tuff fragments in deposits of this and the Bajocian stage signify limited volcanic activity in the vicinity. The Middle Jurassic is represented by condensed red limestones with ferromanganese crusts and nodules, and stromatolites, with some *Bositra* lumachelles. These deposits accumulated slowly on the tops of current-swept seamounts at a depth of a few tens of metres.

The younger Jurassic rocks consist in some localities of radiolarian cherts and marls passing up into coccolith limestones, deposited in basins, and elsewhere of pelagic oosparites signifying persistence of seamount conditions. From Oxfordian times onwards the topographic contrasts of the

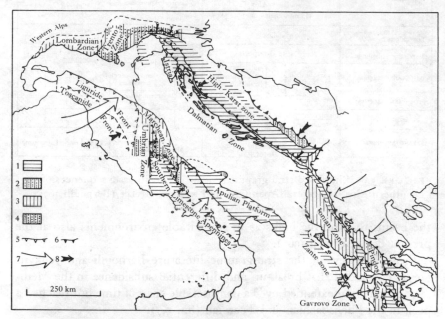

FIGURE 5.9 Alpine tectonic units and Upper Jurassic palaeogeography in the peri-Adriatic region. 1, carbonate platforms; 2, platform margins (reefs, slope deposits); 3, deeper marine troughs; 4, submarine swells, seamounts; 5,6, overthrusts; 7,8, direction of overthrusts (Bernoulli 1971).

sea floor diminished as the basins filled with sediment and the seamounts subsided. By the close of the Jurassic most of the region was covered by a widespread blanket of coccolith ooze.

The peri-Adriatic region. The region embracing the southern Alps, Apennines, western Yugoslavia and Greece is reviewed by Bernoulli (1971). More detailed accounts of the southern Alps are given by Wieden-mayer (1963), Bernoulli (1964), Sturani (1971), Bosellini and Broglio-Loriga (1971) and Castellarin (1972); of the southern Apennines by d'Argenio and Scandone (1971), and of Greece by Bernoulli and Renz (1970). The Bakony Mountains of Hungary exhibit a similar facies sequence and are interpreted in a similar way by Galacz and Vörös (1972).

A simplified stratigraphic scheme is given in fig. 5.8 and the principal late Jurassic palaeogeographic units in fig. 5.9. A clear distinction can be drawn between (i) regions which persisted as carbonate platform into the Cretaceous, which include the southern Apennines and Dalmatian karst country, (ii) submarine swells or seamounts such as the Trento Swell of the Venetian Alps and the Umbrian zone of the north central Apennines, and (iii) relatively deep marine troughs like the Lombardy Alps and the Ionian zone of Greece. Algal, pelletal, oolitic and loferite-type limestones characterise the first category, highly condensed red manganiferous limestones the second, and radiolarites and turbiditic limestones the third. Steep topographic gradients are signified by breccia formations along the platform margins and by slopes on which no sediment accumulated. The basinal sequences lack major stratigraphic gaps and are of the order of hundreds of metres thick. The lack of stromatolites in western Greece or the central Apennines suggests deposition on seamounts below the photic zone. As in Sicily, by late Tithonian times the topographic contrasts had diminished and a relatively uniform thickness of coccolith ooze (Maiolica, Vigla Limestone) was laid down over the whole region, reflecting greatly increased phytoplankton productivity.

Northern Calcareous Alps of Austria. In this region the carbonate platform subsided earlier than elsewhere, because the lowest beds of the Adneterkalk are Sinemurian in age. They are underlain by crinoidal limestones which in turn rest upon Upper Triassic reef or back-reef limestones. Both Jurgan (1969) and Wendt (1969) agreed that no emergence of the Dachstein reefs need be invoked to account for the cessation of coral growth before subsidence was renewed. Nor can Fabricius' (1966) proposal that the sharp upward change from coralline to non-coralline facies signifies a lowering of temperature be considered plausible. The type of facies change in question, if not from reefoid at least from shallow water facies signifying prolific organic production of $CaCO_3$, takes place at different horizons in different places even over limited distances and so hardly supports the notion of regional climatic change. The host of evidence reviewed in this chapter suggests tectonic subsidence as the causative factor. Below a depth of several tens of metres reef corals would have ceased to survive provided the subsidence rate exceeded the rate of upward skeletal growth. Algae presumably could have continued to exist at somewhat greater depths.

Though, as elsewhere in the Mediterranean, Alpine thrusting has rendered it hard to reconstruct detailed Jurassic palaeogeography, the three categories distinguished in the peri-Adriatic region can also be distinguished here. Persistence of carbonate platform facies into the late Jurassic is indicated by the Plassenkalk, which is locally reefoid (Fenninger and Holzer 1970). The Adneterkalk signifies the submarine rise or seamount facies, and the Fleckenkalk and Oberalm Beds basinal facies. In their description of another basinal development near Salzburg, Bernoulli and

Jenkyns (1970) were able to trace an evolution in the composition of turbiditic intercalations reflecting deepening environments on neighbouring swells. Within the Lower Liassic Fleckenkalk these are exclusively echinodermal. Within the overlying red nodular limestones and marls, ranging in age from Middle Lias to Middle Jurassic, pelagic bivalve material is admixed with crinoid ossides. At the top of this rock group only pelagic bivalves occur as resedimented units, while in the overlying radiolarite turbidites disappear owing to the elimination of local topographic contrasts.

6 Facies of the United States Western Interior

Though not known in such detail as in Europe, the Jurassic of the United States Western Interior has been more fully researched than other parts of the world, and the general excellence of exposures in a region of rather dry climate allows the tracing of many individual rock units over huge areas. It affords an excellent example of sedimentation in a shallow epicontinental sea on a craton, with restricted access to the open ocean. Furthermore some of the sedimentary deposits differ from those described in the previous three chapters in giving evidence of extensive aridity.

Our modern knowledge of the United States Jurassic is due primarily to the wide-ranging stratigraphical and palaeontological research of R. W. Imlay, who has published a large number of papers over the last few decades. His 1957 review remains the most useful general introduction to the Western Interior Jurassic. Other important stratigraphic accounts of particular regions are those of Harshbarger *et al.* (1957), Peterson (1957), Imlay (1956, 1967a) and Pipiringos (1968). A valuable series of facies and isopach maps has been published by McKee *et al.* (1956).

PALAEOGEOGRAPHY

The Jurassic sea of the Western Interior occupied, to an extent varying with time, part or all of the states of Montana, Wyoming, North and South Dakota, Idaho, Utah, Colorado and New Mexico (fig. 6.1). It was bounded, at times, by high ground to the west, but sediments were also carried in periodically from the south and east. The highest rates of sedimentation were in a north–south trough in the west, running from the region of the Idaho–Wyoming border to central Utah, with thicknesses ranging up to about 3000 m (disappearance of the Jurassic directly to the west is sudden and due to subsequent uplift and erosion). Eastwards the deposits thin rapidly to less than 200 m, but thicken slightly, to nearly 400 m, in the Williston Basin of eastern Montana, North Dakota and southern Saskatchewan. Facies changes in the sediments, which include a variety of terrigenous clastics, carbonates and evaporites, indicate that

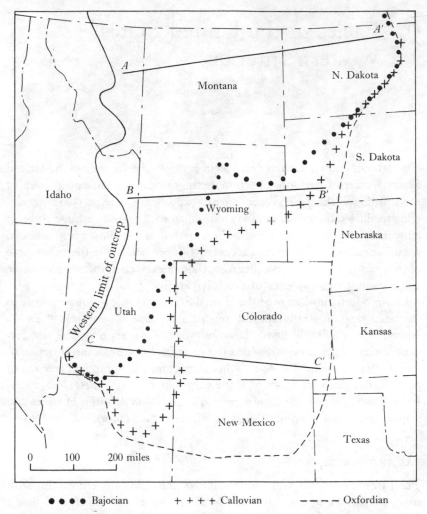

FIGURE 6.1 The approximate southern and eastern limits of the Western Interior sea at different times in the Jurassic. Sections refer to fig. 6.2.

marine influences diminished progressively to the east and south (fig. 6.2).

Within the seaway were a number of islands or shallows, the most important of which were Belt Island in central Montana and the Uncompahgre Swell in west Central Colorado. Though not major sources of sediment, these positive areas influenced the facies to a varying extent. For example, continental red beds do not extend west of the Sweetgrass Arch or Belt Island in Montana. Their primary influence was in reducing sedimen-

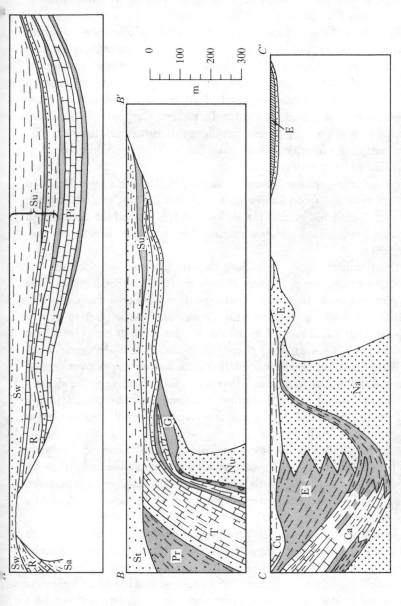

FIGURE 6.2 Sections across the Jurassic of the Western Interior. See fig. 6.1. Ca, Carmel; Cu, Curtis, E, Entrada; G, Gypsum Springs; Na, Navajo; Nu, Nugget; Pi, Piper; Pr, Preuss; R, Rierdon; Sa, Sawtooth; St, Stump; Su, Sundance; Sw, Swift. (Adapted from Imlay 1957.)

tation rates, so that deposits tend to wedge out against them. They are thus comparable to structures like the Armorican or London–Ardennes massifs in the northwest European epicontinental seaway.

The *leitmotif* of Western Interior palaeogeography is a series of advances and retreats of the sea. Four major transgressions of progressively increasing extent can be distinguished (fig. 6.1). The oldest, of Sinemurian age, is recorded in dark, phosphatic shales of the Fernie Group of southeastern British Columbia and probably reached northwestern Montana. Sinemurian deposits are also now known in eastern Oregon and the older concept of a north–south 'Mesocordilleran Geanticline' isolating the northern part of the interior seaway from the Western Cordilleran geosyncline has had to be abandoned.

The second transgression, dated as starting in the high Lower Bajocian by such ammonites as *Stephanoceras* and *Stemmatoceras*, extended much further to the south and east, as far as Wyoming, Utah and the Dakotas; in Montana and northern Wyoming marine Bajocian rests directly on pre-Jurassic rocks.

The third transgression, extending slightly further than the previous ones, has generally in the past been attributed to the Lower Callovian. The sequence of ammonite faunas is: *Arctocephalites (Paracephalites); Arcticoceras (Warrenoceras); Kepplerites* and *Cadoceras*. The latter two genera are unquestionably Callovian. On the basis of his work in East Greenland Callomon (1959) has suggested that *Arctocephalites* and *Arcticoceras* are Upper Bathonian, but precise correlation with Europe has not yet proved possible and the North American subgenera could conceivably be slightly younger in age anyway. No Bathonian ammonites of European type are known from the Western Interior, and this stage is probably represented in part by regressive red beds.

This Upper Bathonian–Lower Callovian transgression is normally marked by a basal shallow water calcareous oolite or sandstone overlain by shales with abundant *Gryphaea nebraskensis*. Locally in southeastern Wyoming these overstep older deposits down to the Triassic, indicating a phase of warping prior to the transgression (Pipiringos 1968).

The fourth and most extensive transgression, which submerged Belt Island for the first time and ranged far to the south and southeast is dated as Lower Oxfordian by the ammonites *Quenstedtoceras* and *Cardioceras*. It was preceded by a widespread phase of regression and local emergence, as marked by the local absence in the north and northeast of higher members of the Lower Sundance formation and by the presence in places of a basal bed with eroded and bored shells and pebbles. One should also note the occurrence at Sykes Mountain, northern Wyoming, of chalcedony nodules in the underlying paper shales of the Hulett beds, apparently signifying a type of caliche and hence emergence (Imlay 1956).

Subsequent to the Lower Oxfordian the sea withdrew definitively from the Western interior, because the fossil evidence clearly indicates a non-marine origin for the overlying Morrison Formation. In late Oxfordian times, however, there was a major transgression around the Gulf of Mexico, but this did not reach northwards far beyond Louisiana. Furthermore, the Upper Jurassic sea appears to have persisted into Tithonian–Volgian times in eastern British Columbia.

A particularly striking feature of the many of the Western Interior deposits is their great lateral persistence. The general excellence of outcrops allows one to trace individual formations and even members and beds over thousands of square miles, over which distance their lithological character may change only slightly. The stratigraphic control is rarely sufficiently good to rule out some diachroneity, but this can at most be only slight. The facts attest conclusively to an environment of low topographic relief and considerable tectonic stability, in sharp contrast to the situation in California and neighbouring states, where distinguishable rock formations cannot be traced over any great distance, and the great sedimentary thicknesses, poor fossil content and subsequent tectonic complications combine to rule out fine stratigraphic analysis. A further pointer to tectonic quiescence in the Western Interior is the extreme paucity of volcanic or volcano-clastic intercalations, again in great contrast to the western geosynclinal region. Furthermore, sharp angular discordances between successive formations are totally absent.

The source areas for the clastic sediments varied with time in dominance. Many of the Lower Jurassic sands have been thought to come from a westerly source, but this straightforward view may have to be modified somewhat, as discussed in the next section. Lateral passage of marine into non-marine deposits and increase in the proportion of sand suggests that during the Middle Jurassic the principal source areas were towards the southwest or southeast. During the Oxfordian, however, they lay to the west. This is attested by the increase in the proportion of sand and scattered pebbles westward and by the local abundance of driftwood, rare elsewhere, in the westernmost Jurassic outcrops in Montana. In their analysis of the Laramide thrust belt along the borders of Idaho and Wyoming, Armstrong and Oriel (1965) argued that uplift prior to the thrusting began in the west early in the late Jurassic, and caused clastics to be shed into the marine basin further east. They also suggested that thrusting might locally have commenced before the end of the Jurassic, but their evidence for this is rather tenuous.

SEDIMENTARY FACIES

One reason why the great lateral persistence of many stratigraphic units has been obscured in the literature is that they have been given different formation names both within and between states, even though the lithology does not necessarily change significantly. Ironically, the term Morrison Formation is used everywhere, although individual subdivisions do not normally persist for any great distance. Fig. 6.3 gives the names in current usage and ties them into the international stages or subsystems. Attention will here be confined to several of the more interesting facies.

Aeolian sandstones

One of the most striking formations of the western Interior Jurassic, outcropping extensively in the Colorado Plateau as spectacularly weathering cliffs, and ranging up to more than 400 m in thickness, is the Navajo Sandstone (Harshbarger *et al.* 1957). This is a white or reddish fine- to medium-grained, almost pure quartz sandstone, exhibiting large-scale trough cross bedding. The Nugget Sandstone further north is correlative and essentially similar in lithology, though with a greater preponderance of red coloration.

Fossils are rare in the Navajo Sandstone, being confined essentially to a few dinosaur footprints, horsetails and obscure ostracods and estheriids; no undoubted marine fossils have been discovered. The base is thought to correspond approximately with the Triassic–Jurassic boundary and the upper boundary may range into the basal Middle Jurassic, i.e. pre-late Lower Bajocian, the age of the first Jurassic marine horizon. While the marked thickening of the Navajo and Nugget westwards, with no significant change of lithology, suggests a predominantly westerly source for the sand, the occurrence of texturally and compositionally similar sands overlying marine Lower Jurassic in western Nevada (Stanley 1971) argues against a continuous 'Mesocordilleran geanticline' acting as a land barrier separating the Western Interior depositional basin from the Western Cordilleran geosyncline. It is highly likely that marine connections with the Pacific existed in a number of places, with scattered large islands acting as sources of sediment.

The spectacular cross bedding of the Navajo and Nugget has long been held to signify a regime of desert sand dunes, and nothing new has emerged to challenge this interpretation seriously. By measuring the maximum dip of the cross bedding, Poole (1964) has attempted to infer the dominant wind directions. For the Nugget Sandstone the dominant direction was determined as northerly and northeasterly, for the Navajo,

FIGURE 6.3 Stratigraphic correlation of the Jurassic formations of the Western Interior. Oblique lines signify stratigraphic gaps.

Stages	Montana	Central Wyoming	S. Dakota and E. Wyoming	S.E: Idaho and adjacent	E. Utah to New Mexico	Subsystems
	Morrison	Morrison	Morrison		Morrison	UPPER JURASSIC
OXFORDIAN	Swift	'Upper Sundance'	Sundance	Stump	Curtis/Todilto	
CALLOVIAN	Rierdon	'Lower Sundance'		Preuss	Entrada	
? BATHONIAN	Sawtooth/Piper	Gypsum Spring	Gypsum Spring	Twin Creek	Carmel	MIDDLE JURASSIC
BAJOCIAN		Nugget		Nugget	Navajo	LOWER JURASSIC

125

northwesterly, except in southwest Utah and northwest Arizona where it was northerly.

The Callovian Entrada Sandstone of Utah, Arizona and New Mexico also consists predominantly of red quartz sandstones, though silty beds also occur. Parts of the sandstone might have been water-laid but much consists of what appears to be normal aeolian dune bedding. Both Poole (1964) and Tanner (1965) have made cross-bedding measurements and determined a predominant northeasterly or north-northeasterly component indicative, they argued, of the principal wind direction (fig. 6.4). The possible palaeoclimatic significance of these data will be considered in chapter 9.

Evaporite bearing deposits

Evaporites occur in a number of non-marine deposits, but most of them occur in the Bajocian Gypsum Spring Formation, the Callovian Preuss Formation and the Oxfordian Todilto Formation, none of which exceed a few tens of metres in thickness.

The Gypsum Spring Formation of Wyoming, Montana and the Dakotas consists of red silty mudstones with gypsum interbedded with platy dolomites and limestones containing a marine fauna of very restricted diversity. Locally gypsiferous beds are absent and replaced by brecciated carbonates indicative of collapse following evaporite solution long subsequent to deposition. The gypsum occurs either as continuous beds or as nodules scattered in the mudstone. One thick gypsum bed, up to nearly 20 m thick, occurs widely throughout northern Wyoming. Examined closely, it is seen to consist of alternations of gypsum and subordinate mudstone layers. It presumably signifies a shallow coastal lagoon (the marine fossils in the associated carbonates signify that the sea was no great distance away), perhaps a subaqueous depression in the supratidal regime, with the interbedded red clastics representing alluvial deposits.

The nodular gypsum appears to be diagenetic, and was probably formed by precipitation following upward or downward movement of ground waters through the muddy matrix sediment. Locally a concentration of such nodules occurs, directly below the marine Bathonian?–Callovian? in a younger red bed unit. Imlay (1956) considered that this might be related to emergence prior to the transgression.

The Preuss Formation of western Wyoming, eastern Idaho and northern Utah mainly comprises a series of red siltstones and sandstones. Marine fossils have been found in the northwest of its area of outcrop, but an important halite unit occurs towards the southeast, as established by drilling and mining. This halite most probably formed subaqueously in a

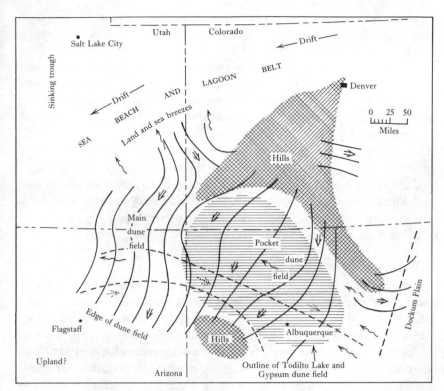

FIGURE 6.4 Late Jurassic, pre-Morrison palaeogeography of the Four Corners region of the Western Interior. Wavy lines show the extent of the early Todilto lake and late Todilto gypsum dune field; solid lines represent the Jurassic wind field; double-shafted arrows fly with the primary-component wind; dotted arrows with the secondary-component wind; wavy-shafted arrows mark the course of flowing water (Tanner 1965).

supratidal depression in an alluvial plain close to the coast, like comparable deposits in Baja California today.

The Todilto Formation of New Mexico, which overlies the Entrada Sandstone, consists of a limestone followed by a gypsum member. The limestone is dark fine-grained and fetid, and is considered by Tanner (1965) to be lacustrine in origin. He believes the gypsum to signify a kind of 'white sands' dune-and-lake deposit. Detailed examination of the limestone reveals the presence of numerous laminations throughout the sequence, made up of alternating limestone, organic and clastic layers (Anderson and Kirkland 1960).

Anderson and Kirkland considered the laminations to be varves. The limestone bands, averaging 0.13 mm in thickness, were held to signify summer precipitates, consequent upon higher temperature evaporation and/or photosynthetic activity. The organic layers, described as sapropel, are more constant and average 8 μm in thickness; they contain subordinate fragments of vascular plants and were regarded as autumn–winter layers resulting from plankton mortality. The third component is detrital quartz sand, intermittent but areally persistent. This was thought to be a winter deposit, both wind- and stream-borne. In the gypsum member, relatively thick gypsum layers are intercalated with the others (fig. 6.5). This interesting interpretation needs to be re-examined in the light of modern work on gypsiferous deposits in the sabkhas of the Persian Gulf.

Taken together with the evidence of dune bedding in the Navajo, Nugget and Entrada sandstones, the presence of widespread and abundant evaporites points clearly to the existence of a warm arid climate in the Western Interior.

Marine Sediments of the Lower Oxfordian

The relatively thin and uniform sediments of the Lower Oxfordian, rarely more than a few tens of metres thick, are an outstanding example of a laterally persistent stratigraphic unit possessing a variety of stratigraphic formation names (Redwater, Swift, Upper Sundance, Stump, Curtis). Unlike the older marine deposits these are everywhere characterised by an abundance of glauconite granules. In the west the Oxfordian is dominantly sandy; eastwards a lower sandy shale unit (Redwater Shale) is succeeded by a sandstone unit (Swift Sandstone).

The Redwater Formation of Wyoming and Montana has been the subject of a detailed investigation by Brenner and Davies (1973). Attention was directed towards the coquinoid sandstone layers which occur within the formation. It is argued that storm-generated currents incised surge channels across pre-existing sand bars and transported shell debris through the channels; bodies of coarse clastics could thus be carried onto mud deposits of the platform beyond. Upon cessation of storm activity, channel lag concentrates of shells, glauconitic pellets and sand were left. The leeward extensions of this material formed so-called storm lag concentrates (fig. 6.6).

The attention given to periodic storm action is welcome, because this factor is often neglected or underestimated in interpretations of shallow marine environments. In the case of the seas of the Western Interior it could have been the dominant agent in creating high energy conditions, because it is difficult to conceive that tidal action could have played a major role in such a shallow sea geographically far removed from the deep ocean.

Brenner and Davies (1974) have also undertaken a more comprehensive

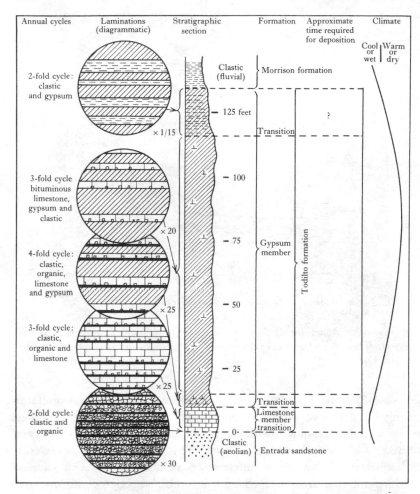

Annual cycles	Laminations (diagrammatic)	Stratigraphic section	Formation	Approximate time required for deposition	Climate

FIGURE 6.5   Minor cycles in the Todilto Formation of New Mexico
(Anderson and Kirkland 1960).

facies analysis of the Oxfordian of the Western Interior. They consider that sedimentation was influenced by three distinct elements. Clastic material was carried into the seaway by a rising westerly source area which was to continue to be of major importance in post-Jurassic time. Secondly, an initial marine transgression was to be followed by an equally widespread regression which was to bring marine conditions to an end, and finally the Belt Island palaeotectonic high profoundly affected sedimentation in Montana. Bordering the shoreline in the west the *nearshore sand facies* contains the coarsest sediments. Eastwards the clastic sediment grain size

129

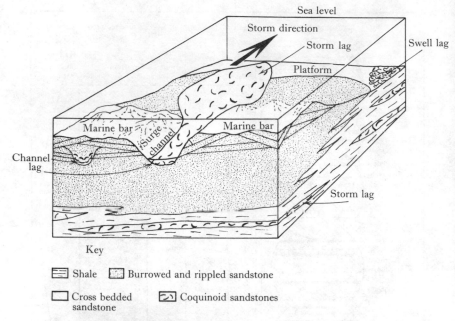

Key

⊟ Shale   ⊡ Burrowed and rippled sandstone

▢ Cross bedded   ⊿ Coquinoid sandstones
  sandstone

FIGURE 6.6 Model for Oxfordian sedimentation in Wyoming and Montana illustrating lateral and vertical relations of marine bars and storm deposits (Brenner and Davies 1973).

decreases gradually and the carbonate content increases concomitantly, giving rise to the *mud facies* followed by the carbonate–clay facies. The latest phase of sedimentation (Swift Formation and its equivalents) is characterised as the *marine bar-sand facies*. This is marked by an extensive orthoquartzitic sand sheet, interpreted as a series of sand bars and muddier interbar areas formed by the combined action of storm-driven, tidal and regional circulation currents. The regional regression is considered to be a consequence of sand progradation from the rising westerly source area rather than epeirogenic uplift over the whole interior seaway. In other words subsidence failed to keep pace with sedimentation.

The interpretation of the youngest Oxfordian deposits as a series of sand bars should not be accepted uncritically. The purported muddier interbar areas are much less obvious in the field than the account by Brenner and Davies might suggest. One is struck rather by the remarkable uniformity at apparently the same stratigraphic horizon of cross-bedded and rippled sandstones with a very shallow water fauna of oysters and trace fossils such as *Rhizocorallium* and *Gyrochorte*, over an area tens of thousands of square miles. Possibly the sands are slightly diachronous and represent shallow subtidal coastal deposits moulded into numerous small dunes. It seems impossible to assess precisely what types of currents were responsible for creating the sedimentary structures.

Non-marine post-Lower Oxfordian sediments

In all regions except the west, where they have been removed by pre-Cretaceous erosion, the youngest Jurassic deposits are grouped in the Morrison Formation (Craig *et al.* 1955). This consists of varicoloured silty shales ranging in colour from purple and red to green and grey, with subordinate cross-bedded sandstone and limestone. Individual beds are hard to trace for long distances and the sand bodies are often very localised. The predominant clay mineral is illite (Moberley 1960), but some bentonitic horizons are found in the west, recording volcanic activity in the Western Cordillera. Workable coal seams occur in Montana and the Dakotas.

Fossils are entirely non-marine. The rich dinosaur faunas, of Wyoming in particular, have long been celebrated, but the unspecialised collector is more likely to find molluscan shells or charophyte oogonia, the latter most commonly in the limestones. Yen (1950) has recognised three bivalve genera, *Unio*, *Vetulonaria* and *Hadrodon*, and nineteen gastropod genera, including *Valvata*, *Viviparus*, *Physa*, *Gyraulus* and *Lymnaea*. The assemblage signifies a late Jurassic, possibly pre-Purbeck age. It is possible, however, especially in Utah, that some of the topmost beds ascribed to the Morrison pass up into the Cretaceous.

The environments signified by the Morrison deposits range from lacustrine and flood plain (shell- and charophyte-bearing limestones and shales) to deltaic and/or fluvial (local sand bodies) and swamp (coal). The occurrence of coal-bearing strata stratigraphically above gypsiferous red beds may well signify a local change to a more humid climate in the north from Middle to late Jurassic time, but this inference should not be drastically extrapolated to the whole of the Western Interior without the finding of stronger evidence.

FAUNAS

In general the macro-invertebrate faunas of the Western Interior are of rather low diversity and not many of the deposits are richly fossiliferous. Bivalves are the only common and reasonably diverse group. Ammonites, gastropods and one or two species of echinoderms occur in moderate abundance sporadically at a variety of stratigraphic levels. Belemnites of the genus *Pachyteuthis* are common in the Oxfordian but almost totally absent elsewhere. The rhynchonellid genus *Kallirhynchia* occurs commonly high in the marine Lower Oxfordian, and one or two species of terebratulids and *Lingula* occur sporadically in the Callovian. Solitary corals, serpulids and bryozoans also occur in places, but little else has been discovered. The microfauna has not been studied adequately but it is known that foraminiferans and ostracods are common in some Callovian shales. Trace fossils include *Gyrochorte*, in platy sandstones of the Callovian and Oxfordian, and *Rhizocorallium, Gyrochorte, Teichichnus* and *Thalassinoides* in the marine Oxfordian sandstones: *Chondrites* also occurs.

The faunas are more diverse towards the north and northwest, as well exemplified by the bivalves. The only Bajocian belemnites and Callovian articulate brachiopods are confined to Montana.

The striking abundance of belemnites in the Lower Oxfordian, as opposed to the earlier marine Jurassic, led Imlay (1957) to suggest that an increased abundance of food might be the factor responsible. It is difficult to see, however, why such an increase, for whatever reason, should not have led to a more general increase in faunal abundance and diversity. In fact the Oxfordian faunas are not more diverse than the earlier ones, and the richest bivalve faunas come from horizons within the Bajocian and Callovian. The presence of ammonites in the two earlier stages also rules out a significant change upwards to more open sea conditions. Rather than invoke purely ecological factors, one should rather take note of the fact that, in northwest Europe, belemnites are appreciably commoner in the Lower Oxfordian than in the preceding two stages, so perhaps the phenomenon is bound up with an evolutionary change within the Boreal Realm.

Another matter of considerable interest is the ecological significance of the bivalve faunas. These have not yet been comprehensively analysed along modern lines, but several points are already clear. Oysters are among the commonest of the bivalves. *Liostrea* usually occurs in high density, low diversity assemblages and is especially characteristic of very shallow water marginal marine environments, while *Gryphaea* is more characteristic of slightly deeper water, more open sea environments in which ammonites also occur. This pattern of distribution can be matched also in the European Jurassic deposits (e.g. Hallam 1971 c). *Liostrea* coquinas are also found in the shallow water Oxfordian sandstones. The genus *Isocyprina* occurs in large numbers in platy dolomitic beds in the Bajocian Gypsum Springs Formation, which suggests that it was a euryhaline form adapted to high salinity 'lagoonal' environments. Limestone beds associated with gypsiferous red beds contain low diversity assemblages lacking normally abundant pectinids, gryphaeas, myas and a variety of other groups, but possessing in a number of localities such genera as *Liostrea*, *Trigonia*, *Tancredia*, *Gervillella* and *Quenstedtia*.

One arrives at the general conclusion that the Western Interior seaway might have at times experienced slightly higher salinities than the ocean, but that these salinities are only likely to have departed appreciably from normal in the more marginal facies towards the south and east.

7 Tectonic history

We may conveniently distinguish two categories of major tectonic and associated igneous activity, associated on the one hand with crustal compression and on the other with crustal tension. The former is characterised by orogenic activity signified by folding and thrusting, metamorphism and plutonism; and also by pronounced marine transgression within small areas, with the local development of major stratigraphic gaps and angular unconformities, and the introduction of wedges of coarse clastic sediments into the rock succession. Such phenomena are characteristically associated with thick sequences of terrigenous clastic and volcanogenic sediments, including flysch, intercalated with lavas and pyroclastics ranging in composition from basic to acid, but generally dominated by andesites, spilites and keratophyres, i.e. a eugeosynclinal association.

Crustal tension is signified by normal faulting and horst-and-graben structure, with basinal subsidence on a large scale; also by fissure eruptions of predominantly basaltic lavas, with associated doleritic intrusives.

Following a review of the principal or better-documented Jurassic compressional and tensional zones around the world that can be recognised on the basic of the above criteria, an attempt will be made to interpret them in terms of plate movements.

Compressional zones

Western Cordillera of North America. The zone extending from British Columbia through Oregon to northern California gives ample evidence of having been a eugeosynclinal belt in the Jurassic (Dickinson 1962). Stratigraphic sequences are often enormously thick, non-sequences and angular unconformities frequent, and evidence of contemporary volcanic activity abundant. This last takes the form of intercalated lavas and tuffs (including pillow lavas), ignimbrites and even the occasional agglomerate; the composition is dominantly augite andesite with subordinate basalt, dacite and rhyolite. Many of the sediments, moreover, are volcanic in

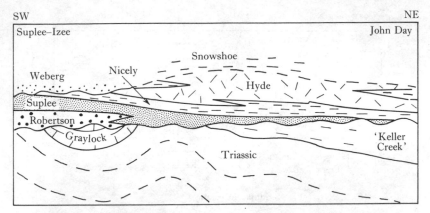

FIGURE 7.1   Lateral variation in the Lower and Middle Jurassic of central
Oregon, illustrating the intra-Liassic unconformity (Hallam 1965).

origin. This is obvious in the case of sandstones and greywackes but is likely
to be equally true of the lutitic beds.

On Queen Charlotte Island, some 900 m of Toarcian sandstones and
argillites rest with angular discordance on Upper Triassic, and are in turn
overlain by up to 3000 m of Bajocian to Callovian volcanics. In the centre
of British Columbia the predominantly volcanic Lower and Middle
Jurassic reach up to as much as 6000 m and are overlain by about 1600 m of
Upper Jurassic greywacke and argillite. In Central Oregon about 6000 m of
Pliensbachian to Callovian deposits, including a thick keratophyric tuff
formation, the Hyde, rest with angular discordance on folded Upper
Triassic and Hettangian (fig. 7.1). Towards the southwest of the state, the
huge thickness of 8000 m has been recorded for a series of Oxfordian and
Kimmeridgian graywackes, argillites and volcanics.

Stratigraphic evidence for the most significant tectonic event, the
Nevadan orogeny, is most clearly displayed in northern California, where
mudstones and greywackes of the Upper Oxfordian–Kimmeridgian Galice
Formation, equivalent to the Mariposa Formation further south, were
folded, metamorphosed and intruded by a granite batholith. Following
erosion, a series of clastic sediments of the Knoxville Formation, beginning with Upper Tithonian, were deposited in the Great Valley region
which, with the Klamath Mountains, is separated from the Franciscan
rocks of the Coast Ranges by a westward-directed thrust, along which lies
a sheet of serpentinised peridotite (Irwin 1964 and fig. 7.2).

The Franciscan rocks are of considerable interest and in the past have
generated a great deal of controversy, which now appears to be substantially resolved. They consist of a great mass of greywacke and mudstone

135

Tectonic history

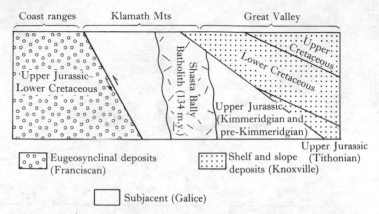

FIGURE 7.2 Diagrammatic west east section across northern California to show relationships of major Jurassic and Cretaceous rock groups. (Adapted from Irwin 1964.)

with subordinate radiolarian chert, limestone and volcanic rocks, and associated isolated masses of serpentinite. Low temperature, high pressure metamorphism is signified by such minerals as glaucophane, lawsonite and aragonite. No normal stratigraphic sequence is recognisable and structures are highly complex. Numerous isolated blocks or olistoliths appear to 'swim' in a foreign matrix, and greywacke units exhibit 'pinch and swell' boudins indicative of local stretching. The dominant features which separate rock units are not bedding but tectonic slide planes. These various features signify that the Franciscan is not a rock-stratigraphic unit but a tectonic mélange (Hsü and Ohrbom 1969).

As regards the age of the Franciscan, three hypotheses have been put forward. The first of these, widely accepted by geologists in the past, including Arkell, is that the rocks pre-date the Lower Cretaceous Knoxville Formation of the Great Valley, from which they are separated by an unconformity. This has been effectively disproved by Bailey *et al.* (1964), who found that the majority of the sparse fossil invertebrates of the Franciscan are late Cretaceous in age. There is no evidence, furthermore, for an unconformity at the base of the Knoxville: the Tithonian *Buchia* Beds in the Great Valley pass up gradationally into Berriasian strata. Bailey *et al.* considered the Franciscan to be coeval with the Great Valley sequence, signifying a eugeosynclinal regime, as opposed to a miogeosynclinal regime, for the latter. This hypothesis is similar to the first in that the Franciscan is considered to post-date the Nevadan orogeny, with the oldest deposits being Tithonian in age. The third view, proposed by Hsü and Ohrbom, is that the Franciscan includes blocks of unfossiliferous pre-Tithonian mixed

136

up with younger rocks. The metamorphism is probably end-Jurassic in age, with minimum radiometric age determinations of about 130 to 150 million years. There is now general agreement with Irwin (1964) that the miogeosynclinal Great Valley sequence has been thrust over a eugeosynclinal Franciscan assemblage, deposited on oceanic crust, at some time in the late Cretaceous or early Tertiary.

Nowhere outside California can a major tectonic episode related to the Nevadan orogeny be so precisely dated (as Lower Tithonian). Radiometric dating indicates that the bulk of the intrusive activity associated with the Sierra Nevada and Idaho batholiths is early and middle Cretaceous in age (Gilluly 1963). A major post-Triassic, pre-Lower Cretaceous episode of regional metamorphism and recumbent folding in the Great Basin is thought probably to relate to the Nevadan orogeny (Armstrong and Hansen 1966). The same is true of the major episode of severe deformation, metamorphism and plutonism in British Columbia, termed the Columbian orogeny and dated as Upper Jurassic (Roddick *et al.* 1966; see Hyndman 1968). In the Canadian Rockies and foothills of eastern British Columbia and Alberta, the basal Cretaceous Blairmore Formation is separated from underlying Palaeozoic and Mesozoic strata by an unconformity of regional significance (Gussow 1960).

Recently Armstrong and Suppe (1973) have reviewed radiometric dates for Mesozoic igneous rocks in Nevada, Utah and California. Magmatism spread widely over the Cordilleran region after about 150 million years ago, with a concentration of activity in the batholith belt from Nevada to California. A major peak of activity is recognisable extending from the late Jurassic into the basal Cretaceous (160–130 million years).

In southern Alaska a series of eugeosynclinal sediments and interbedded andesites, ranging in age from Oxfordian (and possibly Bajocian) into the Cretaceous, have been recognised, and interpreted as signifying an arc–trench system. At some time within the Kimmeridgian–Albian time interval the rocks in the Wrangell Mountains were folded, uplifted and deeply eroded (Berg *et al.* 1972).

The Lower Jurassic Dunlap orogeny of west-central Nevada, to which Arkell attached considerable importance, has been shown by Stanley (1971) to be a local and tectonically insignificant event. The Dunlap Formation is the name given to a group of coarse clastic rocks which overlie Toarcian and older Liassic carbonates and mudstones of the Sunrise Formation. Its components include quartz sandstones derived from the east, which are textually and compositionally similar to the Navajo Sandstone of the Western Interior. This fact is held by Stanley to disprove the existence of a 'Mesocordilleran geosyncline' in Liassic times, a conclusion also reached independently on other grounds (Hallam 1965). The Dunlap also includes locally derived breccias, conglomerates, sandstones and volcanoclastic sediments derived from local andesite flows and pyroclastics. The environment inferred is one of alluvial fan deposition following uplift and volcanism. This new investigation does not support the contention of Ferguson and Muller (1949) that there was thrusting in Liassic times.

Comparing the Lias of Nevada with that in California, the latter is richer in volcanics and much poorer in carbonates, and unconformities are commoner. The existence of lapilli and large-scale cross bedding suggests the possibility of subaerial eruptions on volcanic islands. Clearly the Californian environment was more decidedly eugeosynclinal.

Andes of South America. Locally within northern and central Chile, Lower or Upper Tithonian deposits with basal conglomerates rest with angular discordance on Callovian and older rocks, signifying a late Jurassic fold phase, though elsewhere the discordance is less marked, with Lower Tithonian resting on Kimmeridgian sandstones and conglomerates which themselves signify local uplift (Gerth 1955, Corvalán Diaz 1957, Ruiz *et al.* 1961). In southern Peru, Lower Tithonian beds are claimed to lie with angular discordance on Bajocian (Rüegg 1957). Regional epeirogenic uplift in Mendoza and Neuquen provinces of Argentina, and the Caracoles provinces of Chile, has been dated as late Callovian from the existence of a notable stratigraphic gap below the Oxfordian; locally the presence of an angular unconformity signifies folding or warping (Stipanicic 1966).

In the western part of the cordillera volcanic activity was important through most of the Jurassic (Gerth 1955). The Lower and Middle Jurassic strata of the southern Andes in Argentina and Chile contain abundant interbedded lavas and tuffs. The most important group of volcanics, comprising considerable thicknesses (up to 1000 m) of andesites, keratophyres, quartz porphyries and porphyritic tuffs, is known as the Porphyritic Formation. This is a comprehensive term without precise stratigraphic significance. Thus in northern Chile the age is late Jurassic (probably Tithonian) ranging into Cretaceous, and hence corresponds with the time of significant orogenic events (Harrington 1961); on the other hand, the *Complejo Porfirica* of southeastern Argentina, directly to the east of the Andean chain, comprises three formations, the lower two of which have been dated as approximately Middle Jurassic and the third as Cretaceous (Stipanicic and Reig 1956). The Chon Aikian Formation rests unconformably on sediments dated from fossils as Liassic. Radiometric age determinations of 160 million years for the lavas confirm their inferred Middle Jurassic age (Valencio and Vilas 1970). The overlying Matildian

Formation, consisting of thin quartz porphyry lavas and coarse volcano-clastic sediments containing non-marine *Euestheria* and plant fossils, is thought to be topmost Middle or basal Upper Jurassic.

Most of the huge Andean granitic batholith complex appears to be Cretaceous in age, with radiometric dates grouped around a 100 million year peak, with a possible late Jurassic granite (130 million years) only in one Chilean province (Ruiz *et al.* 1961).

New Zealand. The Mesozoic tectonic history of New Zealand has recently been reviewed by Fleming (1970). A geosyncline became established in the late Palaeozoic and persisted until the late Jurassic. Five structural elements are recognised, and are, from west to east, as follows:

1. A foreland with metamorphic rocks intruded by granite, signifying high temperature, low pressure metamorphism.

2. A margin or hinge line, the so-called Median Tectonic Line, separating zone 1 from zones 3 to 5.

3. A western or marginal belt of moderately fossiliferous sediments, the Honokui facies. These signify rapid and more or less continuous sedimentation in the Triassic and Jurassic but with some local interruptions and non-marine facies in the Middle Jurassic. Lavas are lacking but the sediments contain abundant tuffaceous material.

4. An abrupt facies junction is marked by a narrow zone of ultramafic rocks.

5. The easternmost, structurally complex, belt consists of a thick sequence of poorly fossiliferous deformed sediments including greywackes, argillites and cherts, with associated spilitic lavas and lenses of limestone. No fossils of age between Bajocian and Oxfordian inclusive have been discovered. The variable grades of metamorphism signify relatively low temperatures and high pressures (blue schist facies).

The most significant orogenic episode recognized, which brought the geosyncline to an end, has been termed the Rangitata Orogeny and reached a climax in the Lower Cretaceous, as signified by marine Aptian and younger sediments resting with angular discordance on folded and metamorphosed Upper Jurassic. Precursor movements are thought, however, to have occurred in Middle Jurassic times, marked by uplift in zone 3. The Upper Jurassic appears to have been a time of strong subsidence and metamorphism in zone 5. Radiometric dates for the intrusion of granitic plutons and the uplift and cooling of the metamorphic rocks are concentrated around the middle Cretaceous.

Tectonic history

Eastern Asia. Minato *et al.* (1965) have given a full account of the tectonic history of Japan and the neighbouring land masses. After a general upheaval in the late Triassic the earlier part of the Jurassic was tectonically quiescent in Japan. In the later part of the period local marine transgressions (from the Callovian onwards) and increasing rates of subsidence of the sedimentary basins of the inner, Honshu belt mark the prelude to significant Cretaceous diastrophism and granitic intrusions. In the outer Shimanto and Hidaka belts, facing the Pacific Ocean, a eugeosyncline developed, with much submarine basic volcanicity in the late Jurassic. Tectonism including thrusting commenced before the close of the period and reached a climax in the Cretaceous, with a notable pre-Aptian fold phase and associated granitic intrusions and andesitic vulcanicity. This has been termed the Sakawa Orogeny.

The principal phase of the Yenshan Orogeny of eastern China (Hopei Province) is roughly contemporaneous with the Sakawa Orogeny, and was preceded by a minor phase dated as post-Rhaetian or Lias and pre-Middle Jurassic. Likewise, in Sikhote Alin there is evidence of tectonic movements periodically from Middle Jurassic to Upper Cretaceous, accompanied in their later stages by volcanism and granitic intrusions on a huge scale.

The close of the Jurassic witnessed a major upheaval over the whole of northeastern Siberia, preceded by a considerable intensification of volcanicity in the late Jurassic. Folding and uplift (Kolyma Phase) of the Verkhoyansk geosyncline took place in post-Volgian times, and was accompanied by huge granitic intrusions dated approximately as Lower Cretaceous and, at least in eastern Transbaikalia, by metamorphism (Nalivkin 1960, Beznozov *et al.* 1962).

An approximately end-Jurassic orogeny marked by strong folding and metamorphism is also recognisable in the Phillippines (Gervasio 1967). The sedimentary geosynclinal cycle began in the late Triassic and persisted until the orogeny. In the north Palawan orogenic belt, some 10000 m of Upper Triassic to Upper Jurassic flysch deposits accumulated, with associated spilites and chert.

The Tethyan zone

The Mesozoic tectonic history of the Pamirs compares closely with eastern Asia, in that the phase of maximum intensity occurred at approximately the close of the Jurassic or beginning of the Cretaceous (Nalivkin 1960). In Iran, northeast of the Zagros ranges, a late Jurassic to early Cretaceous fold phase is recorded as the most important tectonic event since Triassic times (Stöcklin 1968).

Locally within the Caucasus and Crimea deposits of Tithonian or approximately Tithonian age rest with angular unconformity on Kimmeridgian or earlier deposits, thereby suggesting to Arkell (1956) a correlation with the Nevadan Orogeny. The most important tectonic phase in these regions, however, was within the Middle Jurassic (the so-called Donetz phase). This phase is clearly shown in the Crimea by the overstep of little-deformed Upper Bajocian strata onto the strongly folded Taurian series of Upper Triassic to Toarcian age (Beznozov *et al.* 1962). Since the Crimea is the type locality of the so-called Cimmerian orogenic phase, it would be better to restrict this term, if it is to be used at all, to a Middle Jurassic episode, rather than apply it as a blanket term for any orogenic movements from Upper Triassic to Lower Cretaceous, as done for instance by Nalivkin (1960).

Intensive volcanicity occurred both in the Caucasus and Crimea throughout most of the period, as marked by great thicknesses of porphyritic andesite, spilite and keratophyre lavas and tuffs (Dzotsenidze 1968). There was particularly significant activity in the Bajocian, with the eruption of a sequence of over 2000 m of volcanic rocks in parts of the Caucasus. According to Dzotsenidze the Great and Little Caucasus mark the sites of independent geosynclinal troughs. The sedimentation rates were considerable, with up to 12000 m of combined Lower and Middle Jurassic recorded locally.

With regard to regions further west, small granitic plutons in Rumania and Greece have been dated as Jurassic, and the Turkish Pontides exhibit evidence of Liassic basalt and andesite volcanism (Smith 1971). Though not obviously bound up with compressional activity, it should be noted that there was a notable phase of Middle Jurassic faulting, uplift (and possibly folding) in the High Tatras of the Carpathians, marked by Bajocian overstepping onto Triassic (Kotanski 1961, Passendorfer 1961). In the High and Middle Atlas of Morocco a minor post-Toarcian, pre-Aalenian phase of warping and uplift preceded major uplift in post-Bathonian, pre-Kimmeridgian times (Choubert and Faure-Muret 1962, du Dresnay 1964a). Uplift of continental hinterland in Tithonian times is also signified by the introduction of flysch deposits into a carbonate regime in a number of circum-Mediterranean countries, such as Yugoslavia, Algeria and Morocco (Bernoulli and Jenkyns 1974) and Bulgaria (Nachev 1966): also parts of the Carpathians and the Caucasus.

Finally, to move to the western side of the Atlantic but still within the Tethyan zone, the conflicting interpretations of the tectonics of Cuba, which Arkell found so confusing, have apparently now been resolved and new evidence rules out a Jurassic orogeny (Krömmelbein 1962).

Tectonic history

Tensional zones

The evidence for regional crustal tension in the circum-Mediterranean countries, as marked by graben subsidence, fault breccias, opening of submarine fissures and minor volcanicity, has been dealt with in chapter 5 and is also well reviewed by Bernoulli and Jenkyns (1974). It need not, therefore, be considered further here. We may instead turn our attention directly to another well-documented region, southern Africa.

The close of the Karroo period in southern Africa was marked by what must have been one of the most spectacular volcanic episodes the Earth has ever seen. At the present day there are approximately 14×10^5 km² of lava outcrop but probably originally some 2×10^6 km² were covered by lavas or affected by the dykes and sills which accompanied the eruptions (Cox 1970). The age of these volcanics appears to range from the late Triassic to the Cretaceous, but it is probably convenient to restrict the term Karroo to the dominant Lower and Middle Jurassic lavas. Radiometric age determinations range from about 200 to 150 million years (Hales 1960, Manton 1968).

The thickness of the lava sequence is greatest in the east, and reaches as much as 9 km in Swaziland and Mozambique. The lavas are mainly tholeiitic basalt with subordinate olivine basalt and rhyolite. Associated intrusive rocks include gabbro, granite, quartz and nepheline syenite, ijolite and carbonatite. The lavas were apparently erupted from fissures, though locally explosive volcanism is indicated by pyroclastics.

Cox found evidence of regional tension not merely in the vast outpourings of lava from fissures but from the widespread normal faulting, warping and subsidence of sedimentary basins. There appears to have been a strong degree of basement control on both the sedimentary troughs of the early part of the Karroo period and the later eruptive zones. Thus in the northern part of southeastern Africa dolerite dykes seem largely confined to, and show a marked parallelism with, the Precambrian orogenic zones. The principle downwarping is associated with the north–south Lebombo Monocline in the east. The sub-Karroo surface descends sharply eastwards under Cretaceous and Tertiary sediments at least 9 km. This was attributed by Cox to crustal extension and 'necking'. Where complete rupture occurred massive volcanism resulted. Where tension was less considerable there was merely a downwarping of sedimentary troughs accompanied by milder volcanism, as in the northern Transvaal.

Middle Jurassic igneous activity indicative of a tensional regime also occurred in eastern Antarctica, because the Ferrar dolerites and basalts have given radiometric ages of 147–163 million years (Evernden and

Richards 1962, McDougall 1963). Likewise the Tasmanian dolerites give an age of 167 million years (Evernden and Richards 1962), suggesting the probability of Jurassic volcanicity here as in Queensland and New South Wales (Arkell 1956). On the other hand the Serra Geral basalts of the Paraná Basin of southern Brazil, formerly presumed to correlate with the Karroo lavas, have been shown to be predominantly Lower Cretaceous, though activity possibly began in the late Jurassic (Amaral *et al.* 1966, McDougall and Rüegg 1966). The basalts of the Kaoka region of South West Africa are correlative with the Serra Geral basalts (Cox 1970).

If these particular igneous rocks have to be removed from the Jurassic record, we can now add the acid and basic lavas and associated granitic ring complexes of northern Nigeria, which have yielded a Middle Jurassic age of approximately 160 million years (Jacobson *et al.* 1964). On the western side of the North Atlantic a coast-parallel basic dyke swarm occurs in southwest Greenland, samples of which have yielded dates of 162 and 138 million years (Watt 1969), indicating a late Lower Jurassic–early Middle Jurassic to late Jurassic age. Radiometric dates embracing the Jurassic have also been recorded from acid and intermediate intrusive rocks in New England (Faul *et al.* 1963, Foland *et al.* 1971). The Cuvo basalts of West Africa are dated as latest Jurassic to early Cretaceous and are associated with graben subsidence along the northwest margins of the continent, which began, however, earlier in the Mesozoic (Reyre *et al.* 1966).

PALAEOTECTONIC RECONSTRUCTIONS

In an investigation into Jurassic plate movements it is necessary to accept a plausible reconstruction of continent–ocean configuration in Triassic times. Such a reconstruction as presented by Smith *et al.* (1973) appears to be satisfactory for this purpose (fig. 7.3). This is based upon the 500 fathom contour least-squares fits of Bullard *et al.* (1965) and Smith and Hallam (1970). There is today virtually no controversy about the fit along the length of the Atlantic or between Australia and Antarctica. The fit of the continental masses flanking the Indian Ocean has provoked more argument because as yet there appear to be few guidelines sufficiently reliable as to be universally acceptable.

The position of Madagascar has posed a particularly thorny problem. Smith and Hallam (1970) followed du Toit in placing the country adjacent to East Africa but Green (1972) and Tarling (1972) have disputed this, preferring a position further south. Geological arguments can be put forward to support either reconstruction, but none appears to be decisive. The former reconstruction of the southern continents will be accepted here (fig. 7.3) but Tarling's alternative is presented for comparison (fig. 7.4). On a world scale the difference is only minor and can for most purposes be disregarded.

There is also considerable uncertainty about the correct placement of the islands and peninsulas of southeast Asia. Whether they should be left adjacent to the Asian continent (Stauffer and Gobbett 1972) or attached to Australia–New Guinea (Ridd 1971, Audley-Charles *et al.* 1972) is another matter that has not yet been resolved to general satisfaction, although the occurrence of a subduction zone on the southern side of Java and considerations of regional geology make the former alternative appear more plausible.

Whatever the dispute about details, all reconstructions of the continents prior to the drifting proposed by Wegener show a supercontinent, Pangaea, substantially separated into a northern (Laurasia) and southern (Gondwana or Gondwanaland) component by an ocean, Tethys, which expanded progressively in width eastwards, and which terminated to the east of what is now Central America. The evidence reviewed earlier in this chapter suggests that the outer margins of this supercontinent lay on or adjacent to subduction zones in the Jurassic, because a eugeosynclinal regime is clearly recognisable along the length of the North and South American Western Cordillera and can be traced less continuously via Western Australia to New Zealand and New Guinea and thence to the eastern borders of Asia. As regards the boundaries of the Tethys, a eugeosynclinal regime presumably indicative of subduction is present in the Caucasus and Crimea. Its possible continuation further west or east, during the Jurassic at any rate, is not firmly established, though it may reasonably be considered probable.

At some time between the Triassic and the present two major plates, Pangaea and the Pacific (perhaps including minor plates), became split into the six recognised today. Abundant evidence exists, though it is beyond the scope of this book to review it here, that this change was substantially accomplished during phases of accelerated and more extensive sea floor spreading in the Cretaceous. This is also consistent with the evidence of increased tectonic and igneous activity in every continent from latest Jurassic times into this period. By the end of the Cretaceous the Americas and India were completely severed from Eurasia and Africa and only Australia remained to be detached from Antarctica in the early Tertiary.

Jurassic plate tectonics is most usefully dealt with in more detail under four headings, the North Atlantic and Arctic, the Western Tethys, Gondwanaland and the Pacific.

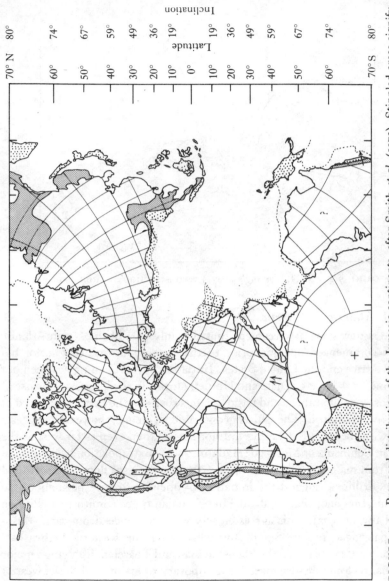

FIGURE 7.3 Proposed distribution of the Jurassic continents, after Smith *et al.* (1973). Stippled areas signify regions deformed in Alpine orogenies whose former position is in consequence relatively uncertain.

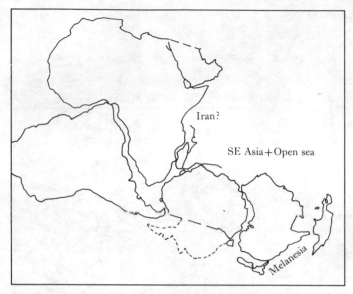

Iran?

SE Asia+Open sea

Melanesia

FIGURE 7.4   D. H. Tarling's (1971) reconstruction of Gondwanaland.

North Atlantic and Arctic

There is now general agreement that the only part of the North Atlantic which commenced opening prior to the Cretaceous was that sector between Africa and the United States. Initiation of this opening has been put by some workers, e.g. Dietz and Holden (1970), in the Triassic, because of the existence of grabens and basic igneous rocks on the eastern seaboard of the United States. The palaeogeographic evidence, however, suggests otherwise, because the Tethys did not extend then even as far west as the Iberian Peninsula or Morocco (Hallam 1971*a*). Extrapolation backwards in time from magnetic anomalies on the Atlantic deep sea floor, and scattered JOIDES drillings, suggest the initiation of opening to have been about 180 million years ago, that is, about Pliensbachian times (Smith 1971, Pitman and Talwani 1972). This agrees closely with inferences from early Mesozoic palaeogeography. Significant collapse of the western Tethys carbonate platform began in the Pliensbachian and Toarcian, implying a major regional tectonic disturbance. Contemporary uplift of land to the west of southern Britain, Portugal and Morocco is evident from the first significant influx of clastic sediments from a present-day oceanic regime (Hallam 1971*a*).

Pitman and Talwani (1972) have attempted a detailed analysis of the kinematics of North Atlantic opening on the basis of magnetic anomaly lineation patterns. Since each anomaly lineation was formed at the mid-oceanic ridge axis, each one represents a former plate margin and is in effect an isochron. Thus by simply fitting together anomaly lineations of the same age but from opposite sides of the ridge, a picture of the ocean and configuration of the bounding continents at that time can be determined. Difficulties are posed in correlating many anomalies, which are often poorly developed. Furthermore the oldest anomaly recognised (anomaly 31) is estimated as being only 72 million years old. Therefore Jurassic movements have to be inferred by extrapolation back in time.

Verification of the technique comes from the parallelism of known fracture zones with the synthetic fracture zones generated from the theoretical model used, from the least-squares 500 fathom continental fit and from JOIDES results.

Pitman and Talwani inferred that Africa separated from North America at a rate of some 4 cm per year from 180 to 81 million years ago, compared with a rate of opening of 2.8 cm per year from 9 million years ago to the present. The Azores–Gibraltar ridge marks the line of shear between Africa and Europe, initially sinistral but changing to dextral in the late Cretaceous when Europe began to separate from North America. Counterclockwise rotation of the Iberian Peninsula took place at some time before the late Cretaceous and the Labrador Sea probably commenced opening in the Jurassic. The Quiet Magnetic Zone boundary is dated at 155 million years (fig. 7.5).

With regard to the Arctic Ocean, it is believed that the Kolyma Block or Massif of northeastern Siberia originally occupied the Canada Basin and was attached to North America. Commencing in the Lower Jurassic, the Canada Basin opened as a result of sea floor spreading, and by the end of the period and into the early Cretaceous the Kolyma Block collided with the east Siberian Shelf and hence caused the creation of the Verkhoyansk Mountains (Herron *et al.* 1974).

Caribbean and Gulf of Mexico. The original Bullard fit of the North Atlantic showed an overlap with the Bahamas Platform, southern Mexico and Central America extending into the region of South America. Freeland and Dietz (1971) eliminated this overlap by suggesting the rotation of the Yucatan, Honduras–Nicaragua and Oaxaca cratons. They inferred an initial break-up of the North Atlantic in the late Triassic, with a concomitant pulling apart of North and South America, causing the Gulf of Mexico to open up as an oceanic sphenochasm progressively during the Jurassic. The Nicaraguan block is thought to have rotated about 2500 km in some 50 million years (a rate of 5 cm per year).

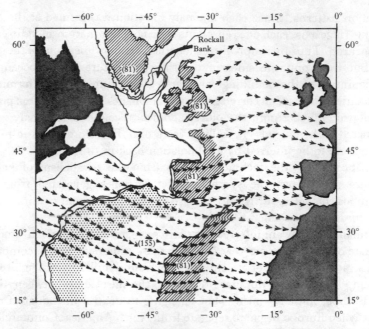

FIGURE 7.5   The relative position of Europe and Africa with respect to North America for the Middle Jurassic (155 m.y.) and late Cretaceous (81 m.y.). Arrows indicate lines of movement. (Adapted from Pitman and Talwani 1972.)

The salt underlying the Gulf of Mexico is considered by Freeland and Dietz to correlate with Jurassic salt on the margining continental areas and to have been laid down in a small oceanic basin with restricted circulation. A more modern view, based on the discovery of apparently supratidal Miocene evaporites in the Mediterranean Basin, would be that the Gulf of Mexico, though truly oceanic, underwent one or more episodes of desiccation during the Jurassic (Hsü 1972).

A quite different interpretation of the Jurassic history of the Gulf of Mexico region has been put forward by Viniegra (1971), on the basis of new borehole information in southeastern Mexico. The salt-bearing Todos Santos Formation is considered to be Oxfordian in age, and hence probably younger than the Louann salt of Louisiana and salt-bearing deposits in Cuba. Jurassic marine deposits wedge out eastwards or southeastwards, suggesting that the sea transgressed from the Pacific. The Gulf of Mexico is considered to have been a cratonic region which was terrestrial in the early Jurassic. Khudoley (in Khudoley and Meyerhoff 1971) also argued that a major land and sediment source occupied the site of the Caribbean in early Jurassic time.

Western Tethys

Any attempt to unravel the complex Alpine tectonics of the Mediterranean region requires an understanding of the so-called ophiolite complexes, which consist of peridotites, pyroxenites, gabbros, spilites and pillow lavas, often associated with radiolarian cherts and other sediments, including pelagic marls and volcanic sandstones. The older idea that these basic and ultrabasic igneous rocks were intruded or extruded into geosynclinal sediments has now been abandoned in favour of the view that they represent thrust slices of old ocean crust or upper mantle (Smith 1971, Bernoulli and Jenkyns 1974). A zone of ophiolites can be followed discontinuously from Oman and Iran through the Taurides of Turkey, the Hellenides and Dinarides, to the Zagreb region, where it is offset by a dextral strike–slip fault (Laubscher 1970). These ophiolites form the uppermost unity of a southwestward-thrust nappe sequence which was emplaced in Cretaceous times during an early phase of the Alpine orogeny. Northwest of the Zagreb line ophiolites occur in the northward-thrust Pennine nappes of the Alps, and are tectonically overlain by the Austro-Alpine thrust sheets with Jurassic of 'Tethyan' (calcareous and siliceous) facies. Significant masses of ophiolites also occur in the Ligurian Apennines.

Jurassic ocean floor can be inferred if dateable sediments of that age can be found to occur in non-tectonic contact with the ophiolites. Characteristically radiolarites lie on pillow lavas, which may pass down into coarser-grained igneous rock. Thus in the Ligurian Apennines some 150 m of carbonate-free thin-bedded radiolarites with intercalated argillite rest on basaltic pillow lavas, with a zone of breccia (ophicalcite) containing radiolarite and lava fragments at the boundary. Upper Jurassic *Calpionella* limestones also occur in association with the North Apennine ophiolites. Perhaps they were deposited on submarine ridge crests or signify a depression of the $CaCO_3$ compensation depth by prolific nanoplankton production (Bernoulli and Jenkyns 1974). Jurassic sediments are also found associated with ophiolites in Yugoslavia and Greece (Smith 1971) and Turkey (Brinkmann 1972).

The next important step in reconstructing Jurassic geography is undertaken by reversing the sense of thrust movements of the Alpine nappes and of strike–slip movement along the Zagros line (Laubscher 1970). Taking into account also the sharp facies contrast between Tethyan calcareous and northern and central European terrigenous clastic facies, the ophiolite-free zones of the Taurides, Hellenides, Dinarides, Apennines, Austro-Alpine nappes and internal Carpathians come to lie on the southern continental margin of the Tethys, while the northern margin is signified by the external zones of the Alpine–Carpathian orogen (Bernoulli and Jenkyns 1974).

Smith (1971) has undertaken another approach to the problem, by attempting a least-squares 500 fathom contour fit of the continental fragments around the Mediterranean. The crustal and upper mantle structure of the deep Mediterranean basins is typically oceanic, though no sea floor spreading magnetic anomaly pattern has been discerned. The basins are thought to represent sphenochasms opened up as a result of the Alpine orogeny, and hence to have nothing to do with the Mesozoic Tethyan ocean. Palaeomagnetic results appear to support the inference of anticlockwise rotation of the Iberian Peninsula, Corsica–Sardinia and Italy, though the results from Turkey are confusing and difficult to interpret in terms of any hypothesis.

These various approaches converge to suggest that the Jurassic Tethys was swallowed up as Africa closed on Europe, presumably along a compressional plate margin from the Alps to the Caucasus. The long-continued but intermittent existence of two parallel geosynclinal troughs in the Caucasus may represent, according to Smith, the approximate position of activated continental margins that formerly lay on opposite sides of the Tethys. This would imply that the ocean disappeared within the intervening Transcaucasian zone, which is marked by a thin or non-existent 'granitic' crustal layer. Smith considered that the Jurassic mid-Atlantic Ridge might possibly have been displaced by a transform fault between North Africa and Spain–Italy, to be continued as a short ridge further east (fig. 7.6). The first Atlantic opening phase, extending from the Lower Jurassic into the Cretaceous, is probably bound up, therefore, with the beginning of the opening of a series of wedge-shaped oceanic areas by the detachment of fragments from the continental area at the western end of the Tethys, with concomitant disintegration of the early Mesozoic carbonate platform.

The position of Iran in this reconstruction, whether it lay nearer or adjacent to the northern or southern side of the Tethys, has remained doubtful. It seems extremely unlikely that it existed as a coherent entity before the Tertiary. The Crush Zone north of the Zagros Mountains marks the line of separation of two geologically contrasted regions (Stöcklin 1968). To the south there existed during the Mesozoic a carbonate platform with no record of major tectonic disturbance, while to the north, in the region embracing the Iranian plateau, the facies is dominantly terrigenous clastic, and there is evidence of more than one major Mesozoic tectonic disturbance. The Crush Zone is also associated with a discontinuous belt of ophiolite-chert coloured mélange, evidently marking the line of ocean closure which, judging by the evidence of Maastrichtian with detrital fragments of the coloured mélange series resting unconformably on the older sediments, was accomplished by the close of the Cretaceous. The existence of other ophiolite zones in central and eastern Iran suggests the possibility of at least one microcontinent (Takin 1971).

As to the extent of ocean separating southern Iran from the rest of the country during the Jurassic, the only clues at present are faunal. Whereas the Zagros Jurassic has faunas with clear affinities to the Ethiopian province at least some of those in east central Iran have strong affinities with northern Europe, as will be discussed in detail in chapter 10. One may therefore tentatively suggest that most of the present country of Iran lay on the northern side of the Tethys.

Dewey *et al.* (1973) have recently adopted a similar approach to that of Smith but their detailed analysis involves more microplates and a more complex, though equally speculative, history of plate movements. One notable difference from the Smith model is that, on the grounds of palaeomagnetic evidence the 'Carnic Plate', including the southern and eastern Alps, is placed on the northern side of the Tethys adjacent to Europe. Furthermore, individual cratonic blocks such as Iberia and Rhodope are thought to have been isolated, at some stage, by ocean, and Turkey might not have existed as a coherent entity in the early Mesozoic. Africa north of the South Atlas Fault, a presumed transform, is considered to have been composed, in the early Jurassic, of more than one microplate, such as the Moroccan and Oran mesetas, separated by narrow strips of oceanic crust.

Movement began in about Toarcian times, coincident with and a necessary consequence of the central Atlantic opening (fig. 7.6). An accreting plate margin with transform segments in the central Atlantic passed into the High Atlas Trough and thence to a triple junction near Sicily. An Alpine branch of this system is recorded by the Piedmont Trough. Iran was separated from Arabia by a narrow oceanic seaway where spreading is thought to have commenced by Triassic times. The initial pattern of plate motion lasted until about Bathonian times, when local changes in plate configuration and relative motion took place. In the late Jurassic (fig. 7.7) the relative motion of Africa and Europe changed slightly and the central Atlantic ridge might have shifted closer to Africa.

The evidence for a series of minor plates, based primarily on the existence of troughs of subsidence without any significant occurrence of ophiolites, is rather tenuous, and the placement of central Iran on the

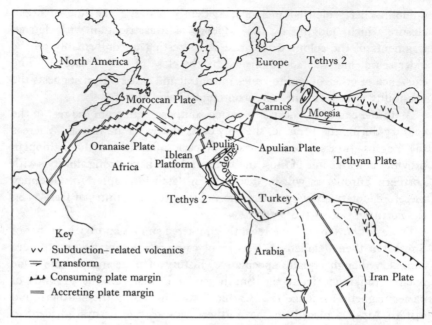

FIGURE 7.6 Proposed Toarcian plate-boundary scheme for the Tethyan
region (Dewey *et al.* 1973).

southern side of the Tethys is apparently not in accord with the faunal and
floral evidence, as noted above.

Gondwanaland

Although Dietz and Holden (1970) have put the commencement of
dispersal of the components of Gondwanaland as early as the Triassic, the
consensus today is that in fact it began well into the Cretaceous. This is
evident from the results of continental palaeomagnetism (Briden 1967,
McElhinny 1970), extrapolation backwards in time of ocean floor magnetic
anomaly patterns, and JOIDES drillings (Maxwell *et al.* 1970, McKenzie
and Sclater 1971, Laughton 1973), faunal distribution (Hallam 1973) and
vulcanicity along the margins of the Atlantic and Indian oceans (Cox
1970). This is not to say, of course, that the initiation of the tensional forces
that were subsequently to separate the continents did not begin earlier.

Cox (1970) has paid particular attention to this problem with regard to
the Karroo volcanics. The Lebombo Monocline and the postulated northeastward projection of the Limpopo Zone are seen as the line of initial
break-up which underwent crustal thinning under tension from Lower
Jurassic into Cretaceous times, with associated volcanicity. The second

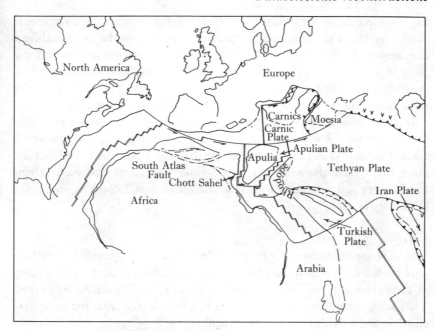

FIGURE 7.7 Proposed Kimmeridgian plate-boundary scheme for the Tethyan region. For explanation of symbols see Fig. 7.6 (Dewey *et al.* 1973).

phase was marked by the opening of the Atlantic during the Cretaceous, associated with the eruption of lavas in southern Brazil and South West Africa, and the third phase by the separation of Madagascar from India at the close of the period, with the associated Deccan eruptions. In the context of this interpretation the zone of marine Jurassic sediments along the margins of East Africa and Madagascar, now known to continue into South Africa (Dingle and Klinger 1971), can be envisaged as marking the site of a shallow and narrow marine gulf that occupied a region of thinning crust which subsided isostatically, more or less along Cox's line *A*. In eastern Tanzania the most significant tensional collapse took place in the Middle Jurassic, as marked by important easterly downfaulting and marine transgression locally onto the Precambrian basement (Kent *et al.* 1971).

On the basis of palaeomagnetic data from Middle Jurassic lavas in southeastern Argentina, Valencio and Vilas (1970) have argued that South America began to separate from Africa as early as the Lower Jurassic. This is hardly consistent with other data suggesting that extensive continental links persisted until Aptian times, when the first Mesozoic marine sediments appeared on the Atlantic continental margins. Moreover, the most recent analysis of magnetic anomalies on the ocean floor seems to discount

Tectonic history

any pre-Cretaceous separation (Larsen and Pitman 1972). If the results are valid they could perhaps signify that drifting apart of the two continents commenced in the south and extended gradually northwards, with complete separation not being effected until the middle of the Cretaceous. According to Dalziel and Cortés (1972) South America is unlikely to have separated from Antarctica before the late Jurassic at the earliest.

It has been widely believed that the Wharton Basin in the eastern Indian Ocean, between Australia and the 90 East Ridge, might contain relatively ancient oceanic crust, but recent JOIDES drillings have established that it is floored by Cretaceous and younger deposits. The oldest deposits overlying basalt occur just west of Australia and are dated as Tithonian–Berriasian (Heirtzler *et al.* 1973).

Pacific

It is now established from JOIDES drillings that only a limited part of the northwest Pacific is underlain by Jurassic ocean floor and we have virtually no direct clue to throw light on Pacific sea floor spreading during the period. That such spreading did in fact occur can be inferred from the eugeosynclinal zones which surround the ocean.

Yeats (1968) has discussed the Californian rocks in this light and concluded that the Franciscan signifies a subduction zone which cannot be related to the westward drift of the North American continent, and which has on the contrary had the effect of eliminating the Mesozoic marginal trench system by low angle thrusting in early Tertiary times. Since that time a new tectonic pattern has developed, associated with the east Pacific Rise. Yeats invoked, instead, spreading of the Pacific floor from the Darwin Rise, which Menard (1964) inferred from evidence of subsidence of the Mid-Pacific atolls and guyots. According to the interpretation of the Franciscan as a tectonic mélange of wide age range, its sediments could have been deposited over a large area of ocean floor before being 'plastered' against the North American continent.

Likewise, Fleming (1970) related the geosynclinal and orogenic phenomena to a subduction zone swallowing up ocean floor spreading from the Darwin Rise to the northeast (New Zealand was subsequently 'rotated' by northeast–southwest movements). Certainly the position of the Darwin Rise as portrayed by Menard fits both interpretations very conveniently (fig. 7.8), but the suggestion of contemporary subduction zones under or adjacent to Japan, Alaska and the Andes virtually boxes the compass and therefore weakens the interpretation. Another serious difficulty has arisen by the failure to discern a magnetic anomaly pattern associated with the Darwin Rise. The results of JOIDES drillings have also cast doubt on the structure, because they fail to provide evidence of *regional* subsidence as opposed to subsidence of independent groups of atolls or guyots. A final point which undermines belief in the Darwin Rise as a Mesozoic spreading ridge is that, according to the tenets of plate tectonics, it should have continued laterally until passing into transform faults, but no such pattern is apparent.

Our best source of knowledge of Mesozoic spreading patterns in the Pacific is the recent paper by Larsen and Chase (1972), who have analysed magnetic lineations in the western north Pacific and correlated them with the results of JOIDES drillings. The authors distinguished three sets of older lineations, termed the Japanese (trend east-northeast), Hawaiian (northwest) and Phoenix (east-northeast), and inferred a Mesozoic pattern of five spreading ridges intersecting at two triple points. The oldest part of the Pacific occurs just east of the Marianas trench and is believed to be early Jurassic in age (JOIDES drillings in this region penetrated Upper Jurassic well above basement). Older crust also occurs on the Magellan Rise, south of the Mid-Pacific mountains, where Upper Jurassic has been found to directly overlie basement. An area of oceanic lithosphere equal to most of the Pacific basin has been subducted beneath Asia, the Americas and probably Antarctica since the early Cretaceous. Quite evidently, therefore, the pre-Cretaceous spreading history of the Pacific can only be guessed at, because the relevant record has been lost under the surrounding continents.

An independent clue as to possible ocean floor subduction on the north-eastern side of the Pacific comes from the analysis of radiometric dates of Mesozoic igneous intrusives in the cordilleran region of North America (Armstrong and Suppe 1973). A late Mesozoic phase of batholith intrusion and vulcanicity (180–75 million years) correlates closely with dates for the metamorphism of the Franciscan (150–70 million years). The pattern of radiometric dates changed dramatically at about 75 million years ago, in the late Cretaceous. Armstrong and Suppe suggested that this difference marks a major change in plate tectonic history, with dip–slip subduction beneath western North America being replaced by strike–slip motion along the newly-created San Andreas Fault.

TECTONICS OF NORTHWEST EUROPE

It is desirable to conclude the chapter with a short account of tectonics in northwest Europe because of the region's importance in Jurassic studies.

One of the best-studied areas tectonically is Lower Saxony in the Federal German Republic, the type area of Stille's 'Saxonic' folding in the

Tectonic history

Mesozoic. The most important disturbance relevant to this review, called the Osterwald phase, is signified by a widespread transgression of the Wealden Serpulite which oversteps rocks as old as Palaeozoic on the Dutch–German border. Accepting the revised correlation of Casey (1963), this signifies warping, uplift and erosion at the close of the Jurassic. The Deister phase, signified by the westward transgression of the Lower Volgian–Tithonian Gigas Beds down to Triassic in the Netherlands (Pannekoek 1956), is of lesser importance (Hoyer 1965).

To a not inconsiderable extent these movements, together with the more significant Cretaceous movements of the same region, may have been due to the diapiric rise of domes of Permian salt (Trusheim 1957). Trusheim has demonstrated that at least some of the local thickness variations and non-sequences in the Jurassic relate to the uplift of such domes, with the concomitant development of peripheral sinks.

Similar conclusions have been reached for the southern North Sea region by Brunstrom and Walmsley (1969). They inferred that important structural features and stratal thickness changes in the Mesozoic and Tertiary have been caused by the movement of Zechstein salt creating swells and basins. One of the most important structures is the northwest–southeast Sole Pit Trough which contains some 1400 m of Jurassic compared with only 600 m on the East Midlands Shelf to the west and even less to the east. It bears a remarkable resemblance to the Gifhorn–East Holstein Trough of Lower Saxony, which presumably might therefore also represent a zone of subsidence due to thinning of the underlying salt beds. The Jurassic was evidently a time of major salt movement and large-scale development of peripheral sinks, but actual piercement structures developed only more recently (fig. 7.9).

Mesozoic and Tertiary salt diapirism is also a notable feature of Portuguese geology and Oertel (1956) has invoked salt movements at depth to account for local uplifts and marine transgressions in the Upper Jurassic of Estremadura. Both in this and neighbouring regions of west-central Portugal there was significant regional uplift of the Iberian Meseta so that in approximately Oxfordian–Kimmeridgian times marine limestones and marls were replaced by a regressive facies of non-marine clastic red beds. Likewise, coarse Kimmeridgian clastics rest non-consequentially on Toarcian in Asturias (northwest Spain), signifying important intra-Jurassic uplift (Dubar and Mouterde 1957). More generally, marine Middle or Upper Jurassic passes upwards into non-marine facies earlier in the western part of the Iberian peninsula than in the east (Querol 1969). The earliest non-marine beds may be charophyte-bearing bird's eye limestones with intercalated polymict conglomerates of age equivalent to the

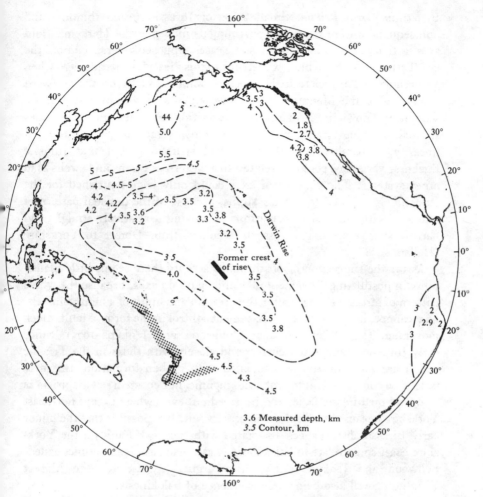

FIGURE 7.8 Relation of the New Zealand Geosyncline to the postulated Darwin Rise (Fleming 1970).

Purbeck Beds, followed by Cretaceous clastic 'Wealden' facies, as in the Aguilar de Campoo district of the eastern Cantabrians (A. M. Sbeta, personal communication). Alternatively, marine Jurassic might pass directly up into 'Wealden' facies, as elsewhere in the eastern Cantabrians, or as in extensive parts of the Iberian ranges of northeast Spain (Beuther *et al.* 1966). Evidently regional uplift of the Iberian peninsula began earlier, and probably continued longer, in the west than in the east.

In England between the two world wars there was a vogue for Jurassic axes of uplift, to which Arkell devoted substantial attention in his *Jurassic*

Tectonic history

System in Great Britain. Small areas of Jurassic stratal thinning and non-sequences were related to posthumous movement of Hercynian fold axes, activity along which had in some cases supposedly persisted until the late Tertiary! A severe blow was struck to this idea a few decades ago when a borehole at Kingsclere in Hampshire, almost exactly on the eastward projection of the Mendip 'Axis', revealed one of the thickest Jurassic sequences known in Britain. Reinvestigation of other purported axes showed that the stratigraphic data employed gave little indication of linearity or parallelism with older fold structures, but that Jurassic thickness variations could be related to a pattern of basins and swells as in other systems. The concept of axes could only have flourished for the Jurassic, as opposed to other systems, because of the unusual pattern of outcrop, with successive formations occupying a relatively straight and narrow strip extending across the country from Dorset to Yorkshire (Hallam 1958).

As for the suggestion of posthumous activity there seems little plausibility in postulating movement of Hercynian fold axes, because the whole pattern of stresses changes after the creation and uplift of a mountain belt. One cannot, however, rule out basement control in the form of fault–block movement (e.g. Whittaker 1972, Sellwood and Jenkyns 1975). Since early Jurassic basins often correspond closely with those in the Triassic which are more obviously associated with graben formation, this may indeed be considered quite probable. Significant Zechstein salt diapirism as a control on Jurassic facies can be ruled out everywhere except northeast Yorkshire because of lack of the necessary salt. It is possible that the minor early Jurassic disturbances associated with the Peak Fault on the Yorkshire coast could relate to *halokinesis*, as salt diapirism is sometimes called. Sellwood and Jenkyns (1975) have argued a case for the Market Weighton Swell also being a consequence of halokinesis.

The most important phases of warping, uplift and erosion in England were, firstly, in the early Middle Jurassic, when regional uplift led to regression in the northeast and the creation of an extremely shallow Bahamian-type platform in the southwest. Post-Toarcian, pre-Aalenian movements occurred in Yorkshire, and intra-Bajocian movements in the Cotswolds (roughly contemporaneous disturbances are recorded from the Ardennes). Secondly, an important phase of regional uplift in the Middle Volgian has been inferred by Casey (1971) on the basis of an extensive horizon of black chert pebbles ('lydites') and worn and phosphatised fossils derived from the underlying Kimmeridge Clay (fig. 2.2). This regional unconformity may be dated on the reasonable assumption that the *Pavlovia rotunda* nodules of the Upper Kimeridge Clay correlate with the base of the Middle Volgian of the Russian Platform. The lydites appear to be derived from Lower Carboniferous limestone, which must have been extensively exposed in northern England.

It is noteworthy also that in Scotland no Jurassic rocks younger in age than Lower Kimmeridgian have been discovered, but since there is a major stratigraphic gap the regional uplift could have taken place in the Cretaceous.

Major Lower Kimmeridgian activity along a submarine fault scarp in northeast Scotland is signified by the celebrated boulder beds of Helmsdale, which consist of blocks of Old Red Sandstone which have tumbled into a muddy marine basin (Bailey and Weir 1932, Crowell 1961). Equally spectacular sedimentary breccias related to normal faulting parallel to the coast occur in the Upper Volgian or Berriasian Rigi Series of central East Greenland, marking the first tectonic disturbance there of major significance in the Mesozoic succession (Donovan 1957, Surlyk *et al.* 1973).

There is more than a possibility that the intensification of tectonic activity towards the end of the Jurassic correlates with an increase in halokinesis, as though regional tectonic disturbances triggered off increased salt diapirism; this is suggested by the works previously cited and from similar evidence in western Morocco (Société chérifienne des petroles 1966).

This general intensification of tectonism from the late Jurassic into the Cretaceous could be a local expression of the world-wide changes that have been discussed earlier in this chapter. At any rate the evidence is strong for a synchroneity of tectonic phases over an extensive area of the North Atlantic, because events deducible from the sedimentary sequence in East Greenland bear a close resemblance, in a number of respects, to events in northwest Europe (Surlyk *et al.* 1973). Thus the sea deepened or spread in both regions in early Pliensbachian, middle Toarcian and late Oxfordian times, and there was a minor phase of uplift or regression in the late Lower Jurassic, followed by a more important phase towards the close of the period. According to Surlyk *et al.*, tectonic activity increased notably in Kimmeridgian–Volgian times, culminating in spectacular block-faulting and tilting. This situation is directly comparable to northeastern Scotland, where the evidence of the Helmsdale boulder beds points to large-scale normal faulting and the collapse of a trough to the east. It is now established that a zone of tilted horst-and-graben structures underlies the North Sea between Scotland and Scandinavia, and may extend a considerable distance northwards.

An attempt can tentatively be made to relate major tectonic events in the British area to phases of opening of the North Atlantic as determined by Pitman and Talwani (Hallam 1972*b*).

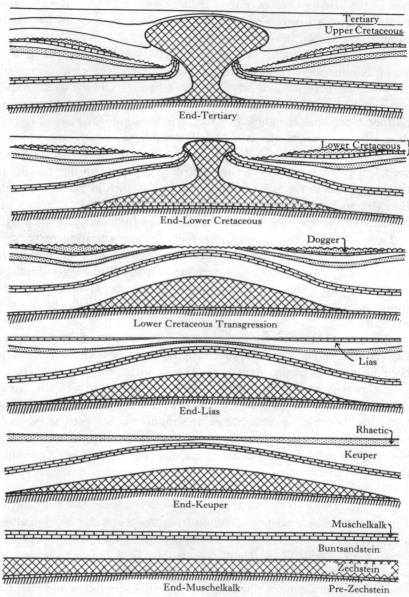

FIGURE 7.9 Salt-dome evolution during the Mesozoic and Tertiary of northwest Germany. (Adapted from Trusheim 1957.)

Major geomorphic features seem to have been initiated fairly shortly after the Hercynian orogeny, so that basins in, for example, the North Sea, English Channel, Irish Sea and the Minch may be considered to date from the early Mesozoic.

Subsequently, at some time between late Jurassic and late Cretaceous, and most probably quite early during this interval, much of western Britain appears to have been uplifted above sea level and subjected to erosion, with the maximum uplift taking place in the Irish Sea region. This was a time of widespread tensional faulting, with collapse of basins such as the Moray Basin and the uplift of adjacent horsts, among which western Britain can probably be included.

The early Mesozoic graben subsidence appears to correlate with the first phase of Atlantic opening and seems to signify tensional effects extending northwards from the principal zone of rifting tens of millions of years prior to Atlantic spreading at this latitude. The second phase followed a period of significant epeirogenic movements involving the creation of horsts such as western Britain and grabens such as the Shetland Trough. Widespread subsidence followed as spreading commenced, and late Cretaceous deposits mantle older rocks unconformably, both on land and the continental shelf.

It is now known that a phase of updoming in the central and northern North Sea took place in Middle Jurassic times, and was responsible for the regional regression apparent from a study of the north west European facies. It was accompanied by trilete rifting, i.e. three rift systems meeting at a spot close to the Forties Oilfield, and, in the Bathonian, by considerable volcanicity (Sellwood and Hallam 1974). The rifting constituted what has been termed a 'Failed Arms System' and was not followed by sea floor spreading. Instead the grabens continued to subside throughout the later Jurassic, with marginal faulting especially in the northern North Sea or Viking Graben; the porous Jurassic sandstones accumulated in this graben have proved to be very important reservoir rocks for petroleum (Howitt 1974). A wealth of information on North Sea geology was brought for the first time into the public domain at a conference held in London in November 1974, the proceedings of which are due to be published in 1975 by Applied Science Publishers. One of the outstanding facts to emerge is that older deposits of various ages are mantled by thick, bituminous Kimmeridge Clay, which is downfaulted in the grabens against porous sandstones. It seems to be by far the likeliest source rock for the oil which has accumulated, for instance, in non-marine basal and Middle Jurassic sandstones in the Brent field, east of the Shetlands, and the marine Upper Jurassic sandstones of the Piper field, east of the Moray Firth.

8 Changes of sea level

SEDIMENTARY CYCLES

Over half a century ago the German geologist Klüpfel (1917) undertook an analysis of sedimentary cycles in the Lower and Middle Jurassic of Lorraine, in which a sequence of ammonite-rich shales and marls alternate with calcareous sediments exhibiting faunal and sedimentary features suggestive of deposition in shallower water; the latter were often capped by one or more successive hardgrounds. Klüpfel took these to signify emergence above sea level and inferred local relatively rapid epeirogenic subsidence followed by renewed but more gradual shallowing as the zone of marine deposition filled up once more with sediment. In his account of the British Jurassic Arkell (1933) attempted a subdivision of the succession in different regions into cycles of clay, sandstone and limestone, representing a progressively shallowing sequence again attributed to epeirogenic oscillations.

A survey of the Lias in northwest Europe indicated that a series of non-sequences, or horizons of condensed or relatively shallow water deposition, alternating with deposits signifying more continuous and presumed deeper water sedimentation, extended right across the area (Hallam 1961). Particularly significant was the widespread change from the Pliensbachian to the Toarcian, subsequently analysed in more detail (Hallam 1967c), where a series of shallow water ironstones, sandstones and limestones pass within the limits of only one or two ammonite zones into deeper water shales and marls. The intriguing fact emerged that these various changes, indicative or at least suggestive of variations in depth of sea, were independent of a regional pattern of basins and swells. Since basins and swells are most reasonably interpreted in the usual way, as signifying differential epeirogenic subsidence, a problem arose concerning the laterally more extensive changes. Was one to invoke a second type of epeirogenic subsidence, operating more or less uniformly over areas of hundreds of thousands of square kilometres and superimposed on the regional pattern, or was there some more plausible explanation?

The only reasonable alternative is that the extensive changes resulted from world-wide or eustatic changes of sea level. To test this an attempt was made to correlate changes apparently indicating deepening of the sea with marine transgressions in the different continents, as signified where continental sediments pass up into marine sediments or where the latter overlie older deposits with a notable stratigraphic gap. Only in this way can one firmly exclude the possibility of local tectonic control. The results of a preliminary survey suggested that there was such a correlation for at least five levels in the Jurassic, the Lower Pliensbachian, Lower Toarcian, Lower Bajocian, Lower Callovian and Lower Kimmeridgian.

Two questions arise. Does detailed facies analysis bear out the suggestion that Jurassic sedimentary cycles relate to changing water depth, and does the asymmetry characteristic of many such cycles necessarily reflect asymmetry in the controlling mechanism, as originally inferred by Klüpfel?

In his comprehensive and thorough study of part of the Lower Lias in Britain Sellwood (1970) recognised three types of minor sedimentary and faunal cycle developed in clastic and calcareous sequences. In each case analysis of the sediments and fauna pointed to increases in environmental energy towards the tops of the cycle, which were most plausibly related to shallowing of the sea. Sellwood's interpretation of his type 3 cycle, developed in the Jamesoni Zone of Yorkshire and Mull, is given in fig. 8.1 to illustrate this point. Sellwood suggested a eustatic superimposed on a local epeirogenic control. Tectonic/compactional subsidence combined with a eustatic rise of sea level would lead to relatively rapid deepening. If subsidence were intermittent and eustatic changes of a few metres superimposed, then cycles should have resulted naturally in shallow extensive epeiric seas whose sedimentation could be expected to respond in quite a subtle way to small-scale effects.

Purser's (1969) work on Middle Jurassic limestone–marl sequences of the eastern Paris Basin throws doubt on Klüpfel's interpretation that the limestone hardgrounds signify emergence and hence represent the shallowest part of the cycles. On the contrary, as discussed in chapter 4, they appear to have formed subaqueously under conditions of slow sedimentation at a greater depth than limestones beneath them. Purser accepted eustatic control as a likely interpretation. In this case, slow shallowing followed by rapid deepening need not be invoked; both positive and negative eustatic changes could have taken place at equal speed.

If the southern English Upper Oxfordian cycles discussed by Talbot (1973) and considered in chapter 4 are also eustatically controlled, which is suggested at least tentatively by the apparently close correlation of one major transgression with the French Jura, then they could have resulted

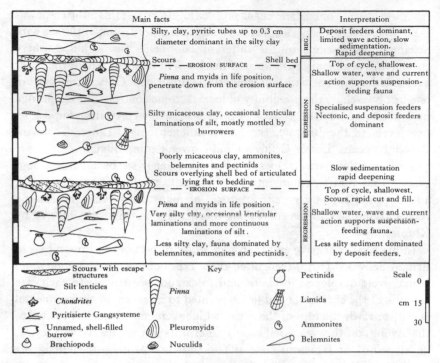

Main facts		Interpretation
	Silty, clay, pyritic tubes up to 0.3 cm diameter dominant in the silty clay	**REG.** Deposit feeders dominant, limited wave action, slow sedimentation. Rapid deepening
	Scours —EROSION SURFACE— Shell bed	Top of cycle, shallowest. Shallow water, wave and current action supports suspension-feeding fauna
	Pinna and myids in life position, penetrate down from the erosion surface	
	Silty micaceous clay, occasional lenticular laminations of silt, mostly mottled by burrowers	**REGRESSION** Specialised suspension feeders Nectonic, and deposit feeders dominant
	Poorly micaceous clay, ammonites, belemnites and pectinids Scours overlying shell bed of articulated lying flat to bedding —EROSION SURFACE—	Slow sedimentation rapid deepening
		Top of cycle, shallowest. Scours, rapid cut and fill.
	Pinna and myids in life position. Very silty clay, occasional lenticular laminations and more continuous laminations of silt.	**REGRESSION** Shallow water, wave and current action supports suspension-feeding fauna.
	Less silty clay, fauna dominated by belemnites, ammonites and pectinids.	Less silty sediment dominated by deposit feeders.

Key

Scours 'with escape' structures			Pectinids	Scale	0
Silt lenticles		*Pinna*			
Chondrites			Limids	cm	15
Pyritisierte Gangsysteme					
Unnamed, shell-filled burrow	Pleuromyids		Ammonites		30
Brachiopods	Nuculids		Belemnites		

FIGURE 8.1 Diagrammatic section through Sellwood's type 3 cycle in the British Liassic (Sellwood 1970).

from intermittent eustatic rise alternating with periods of basin filling by clastic sediments during times of stillstand, with more or less uniform subsidence. Whether such an interpretation is more generally true obviously depends on the sensitivity of sedimentation to slight changes in water depth, and each individual case must be examined on its merits.

More generally, it has been pointed out by Duff *et al.* (1967) that asymmetrical sedimentary cycles of various types might be expressions of symmetrical eustatic controls. This asymmetry has the common characteristic that the transition to the presumed deeper water or more offshore facies is relatively sharp, and such evidently transgressive deposits frequently bear evidence of relatively slow deposition. As discussed in detail in chapter 6 of their book, positive and negative movements of sea level of similar speeds may act in such a way that the transgressive parts of eustatically controlled cycles will often be more condensed than the regressive parts.

Another interpretation of clay–sand–condensed limestone sequences in regions of epeiric sea sedimentation has recently been put forward by

Sellwood and Jenkyns (1975), citing examples from the Pliensbachian-to-Bajocian succession in southern England. In their view, and in contrast to that of Talbot (1973), the limestones mark the culmination of a shallowing-upward sequence. Sedimentation is believed to have been controlled by faulting in the basement, which may have been related to rifting and spreading in the Atlantic and western Tethys, and is not therefore necessarily bound up with sea level changes.

The major limestone units cited, however, such as the Toarcian part of the Junction Bed and the Aalenian–Bajocian in Dorset, can be shown to correlate with transgressive deposits elsewhere, which suggests that Talbot's model for the Corallian (see chapter 4) may be the more appropriate one in this case as well. Furthermore, although the fault-controlled regional subsidence model is an interesting one worthy of serious consideration for some cycles, it is questionable whether it can be applied to those cyclic changes indicative of varying depth of sea which can be traced for hundreds of thousands of square miles.

MARINE TRANSGRESSIONS AND REGRESSIONS

Another approach to the question of changing sea level is to plot the known or inferred original distribution of marine sediments on an equal-area projection of the continents for successive stages of the Jurassic and assess the areal percentage covered by sea at different times. An attempt to do this was made by Hallam (1969b). (Figs. 8.2 to 8.6 use, however, the Smith *et al.* reconstruction as it is a more realistic portrayal of the Jurassic world, though not, of course, an equal-area projection.) Because the change from one stage to the next is usually insignificant it was considered sufficient to plot data from only five well-spaced stages and specify in the text to what extent the unillustrated stages differ.

A number of problems in applying this technique have to be faced. The quality and amount of stratigraphic data vary greatly from region to region – it is especially unsatisfactory, for instance, in southern and eastern Asia. 'It is impossible on the small scale attempted to plot only the proven deposits. This would anyhow defeat the object of the exercise because patchy distribution very often signifies removal by erosion at a later stage. As in all palaeogeographic reconstructions, a certain amount of reasonable extrapolation is necessary, thereby raising the familiar question of whether the absence of deposits from a given area signifies non-deposition or subsequent uplift and erosion. Hopefully, underestimation of former areal extent due to this latter factor may cancel out, to some extent, errors resulting from overestimation based on palaeogeographic inference. Be-

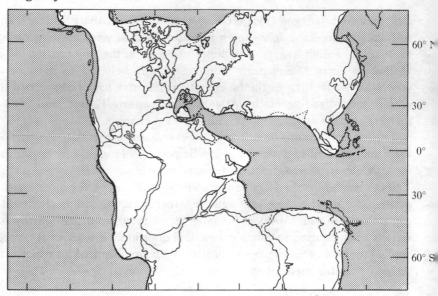

60° N

30°

0°

30°

60° S

FIGURE 8.2   Approximate distribution of land and sea (stippled) in the
Hettangian; small islands excluded.

cause of these various uncertainties the quantitative data presented below
must be treated as very approximate.

The regional references used are given in Hallam (1969*b*).

Hettangian: fig. 8.2. During this stage the areal extent of sea was
minimal, and continental sediments are much more widespread than
marine. The distribution of marine deposits is generally similar to the
Rhaetian, on which they usually rest concordantly, but locally they
overlap it to a variable extent, as for instance in South Wales, the Inner
Hebrides of Scotland and the Ardennes. There are no apparent indications of regression with respect to the Rhaetian.

Sinemurian. A widespread early Sinemurian transgression affected the
North American continent, with the sea spreading, for the first time in the
Jurassic, beyond the Cordilleran geosyncline into Alberta; and also for
the first time into northwest and east-central Mexico. A more or less progressive transgression also seems to have occurred in parts of the Andes,
but details are lacking. The lower Sinemurian marks the time of an important transgression in northern Europe, as indicated in southern Sweden and
Scotland. A progressive transgression also took place over the margins of
Palaeozoic massifs in western Europe, e.g. the Ardennes and Massif

Central. Except for limited areas in Andalusia and the Moroccan Rif, the first ammonite-bearing strata in the Iberian Peninsula and North Africa are Sinemurian and generally overlie lagoonal dolomites and limestones of presumed Hettangian or early Sinemurian age. No notable regression with respect to the Hettangian is known, except perhaps in the Toroya region of Japan.

Pliensbachian. In Lower Pliensbachian times an important transgression affected land currently bordering the North Atlantic; its effects are recorded in East Greenland, northeast Scotland, southern Sweden, Bornholm, the Polish lowlands, the borders of the London–Ardennes and Brittany Massifs and the Massif Central, the northern Pyrenees and Moroccan High Atlas. Locally the transgressive beds begin with the uppermost part of the Sinemurian. During the stage regression took place in Mexico and Alberta. During the Upper Pliensbachian a major transgression took place in eastern Asia, with the sea spreading over continental deposits in western Honshu, Japan, and extending as far inland as the Vilui depression and Upper Amur Basin in Siberia. Transgressive beds of this age, lapping against ancient massifs, are also known in Georgia in the USSR, central Iran and the Moroccan High Atlas, Normandy, France, and the Mendips, England. Further transgressive beds are known in Oregon and Alberta. Regression with respect to the Lower Pliensbachian occurred in East Greenland, the Polish Lowland and the southern border of the Ardennes Massif.

Toarcian: fig. 8.3. Transgression during the Toarcian appears to have been more or less progressive at least into the middle of the stage, following a possible end-Domerian regression. In Lower Toarcian times the sea spread widely in the newly-formed 'Trans-Erythrean Trough', with deposits in Arabia, Baluchistan, East Africa and Madagascar; also in the Donetz Basin, USSR. In Middle Toarcian (Bifrons Zone) times the sea returned to Alberta and East Greenland and covered part of the Canadian Arctic islands and Spitsbergen for the first time in the Jurassic. The Bifrons Zone is also widely transgressive in the Mediterranean region, resting non-sequentially on Pliensbachian, and in central Iran. Toarcian regressive deposits are very subordinate, e.g. in western Morocco. Late Toarcian influx of coarser clastic deposits, and initiation of erosion in parts of England, Portugal, the borders of the Ardennes Massif, the Moroccan High Atlas and Honshu, Japan, can be readily related to local tectonic uplift.

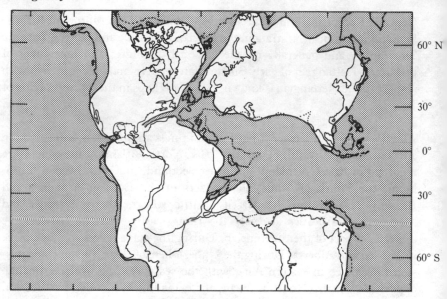

FIGURE 8.3   Approximate distribution of land and sea (stippled) in the
Toarcian; small islands excluded.

Aalenian. The pattern is essentially similar to that of the Toarcian. The sea
returned to the Polish lowland and spread again, following local tectonic
uplift, to northeast England, the eastern High Atlas and Middle Atlas of
Morocco, and Japan. Regression took place in East Greenland, Spitsbergen and Yorkshire (post-Murchisonae Zone), and perhaps also in New
Zealand and parts of East Africa.

Bajocian: fig. 8.4. In the Lower Bajocian (= Middle Bajocian of Arkell
1956) the sea penetrated for the first time into the Western Interior of the
United States, Western Australia and New Guinea. Other areas whose beds
of this age are transgressive include the southern Andes, the Torinosu
region of Japan (?), the Ardennes and northeast England (Discites and
Humphriesianum zones). Upper Bajocian deposits are transgressive in the
Polish lowlands, Georgia and Uzbekistan (USSR), the Balkan Mountains
and, probably, Greenland and the Canadian Arctic.

The Bajocian is regressive with respect to the Aalenian and Liassic in the
Vilui Depression and eastern Transbaikal region of eastern Siberia, where
marine clastics pass up into continental coal-bearing beds, and in the
Kuruma Basin of Japan. Thick regressive beds of coal measure facies are
intercalated with marine deposits in northeast England and northern Iran.
In the northeastern USSR, east of the Vilui Depression, no Bajocian or

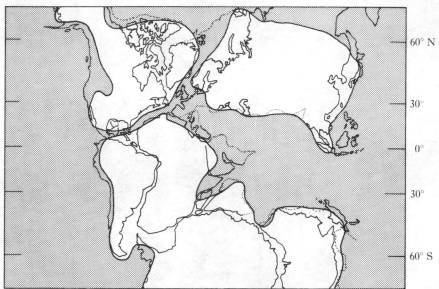

FIGURE 8.4 Approximate distribution of land and sea (stippled) in the
Bajocian; small islands excluded.

Lower Bathonian ammonites have been found but the succession of
Inoceramus species suggests the presence of a more or less continuous
sequence.

Bathonian. The Lower Bathonian marks the time of the first major
reversal in the trend of more or less progressive transgression in the earlier
Jurassic. Almost nowhere did the sea extend further than the underlying
Bajocian, whereas regression was widespread, as stressed by Arkell (1956),
e.g. in the Western Interior of the USA, Scotland, the Caucasus, central
Iran, North Africa and probably at least part of the Andean zone. On the
other hand, transgression on a large scale was renewed in the Upper
Bathonian and is clearly recorded in many parts of the world, e.g. the
Western Interior of the USA, the Canadian Arctic, the Polish lowland, the
margins of the London–Ardennes Massif, north-central Siberia, Tunisia,
Egypt, East Africa, Afghanistan, Pakistan and Burma.

Callovian. The familiar 'Callovian transgression' appears in effect to be a
continuation of the preceding Upper Bathonian transgression and was more
or less progressive at least until the middle of the stage. The effects are
well-marked over a vast area, most notably the Russian Platform and West
Siberian Depression, with a direct marine link being established for the first

169

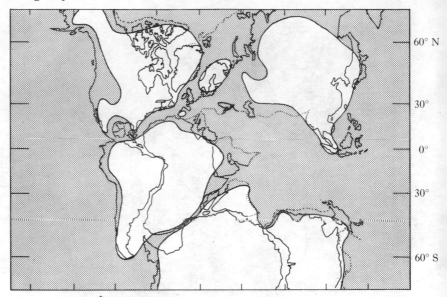

FIGURE 8.5 Approximate distribution of land and sea (stippled) in the Oxfordian; small islands excluded.

time between north and south. The sea spread for the first time into the Saharan zone of Africa (southern Tunisia), and southwards in the Western Interior of the USA. The Kolyma Massif and North Siberian islands were apparently submerged for the first time; other areas of transgressive Callovian include western Morocco, Japan (?), Mexico and (probably) California.

Nowhere is the early Callovian known positively to be regressive with respect to the underlying deposits, but a post-Lower Callovian regression apparently took place in a number of regions such as the Western Interior of the USA and Canada, Greenland, the Canadian Arctic (probably), New Zealand and New Guinea.

Oxfordian: fig. 8.5. The pattern compares closely with that of the Callovian and by the latter part of the stage the seas achieved their maximum extent in the Jurassic. In the Upper Oxfordian the sea penetrated deep into the Gulf Coast region of the USA and neighbouring Mexico and Cuba, while in the Lower Oxfordian the Western Interior sea covered a larger area than at any time previously. An Upper Oxfordian transgression is also recorded in Greenland, the southern Andes, the Polish lowland, Turkmenia, New Guinea and the Canning Basin of Western Australia. Indications of contemporaneous regression are slight, though

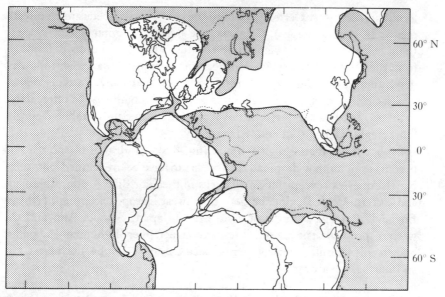

FIGURE 8.6   Approximate distribution of land and sea (stippled) in the late
Tithonian–Volgian; small islands excluded.

some occur in the western part of the Iberian Peninsula. No marine
post-Lower Oxfordian deposits have been proved in the American Western Interior but there is no proof that the continental deposits which
overlie the youngest ammonite-bearing strata begin with Upper
Oxfordian.

Kimmeridgian. The distribution of sea and land appears to have been
substantially similar to the Upper Oxfordian. Transgressive beds occur in
the Gulf Coast region, northern Germany, Holland, northwest Spain,
northeast Morocco, west Algeria, Arabia, Crimea, Azerbaijan, Tanzania
and New Zealand. At least some of the transgression is related to local
tectonics, e.g. in western Europe and North Africa, as indicated in the last
chapter. Regression set in after the early Kimmeridgian in a number of
regions, e.g. the southern and central Andes, southern Tunisia, western
Morocco, Arabia, Georgia, Uzbekistan and possibly Pakistan and New
Guinea.

Tithonian–Volgian: fig. 8.6. In the early Tithonian–Volgian a widespread transgression occurred throughout the Andes (locally the basal
transgressive beds are Upper Tithonian), with the sea extending for the first
time into the eastern Andes of Colombia and into southern Patagonia; also

locally in northwest Germany (where the transgressive beds are manifestly related to local tectonics), Calabria, the Balkan Mountains and northern Caucasus. At about the same time the Jurassic sea withdrew definitively from the Gulf Coast, western Morocco, northeastern Spain and, following deposition of the Gigas Beds, from northwest Germany. Partially regressive facies, with lagoonal evaporites and continental clastics intercalated with marine beds, are developed over wide areas in Saudi Arabia, Georgia, Turkmenia, Uzbekistan and Tanzania.

By late Tithonian–Volgian times (fig. 8.6) withdrawal of the sea had become even more widespread. It appears to have been excluded from all of North America (except for small areas in the Canadian Arctic, California and eastern Mexico), Spitsbergen, northwest Europe excepting northeast England, Portugal, northeast Asia except Japan, Sikhote Alin and the western pre-Okhotsk region. Transgressive Upper Tithonian–Volgian is known only in limited areas, e.g. northern California, Cuba, northern Chile and northeastern England.

Accepting that the method must necessarily give only approximate results, an attempt has been made to plot graphically the area of the continents covered by sea at various times throughout the Jurassic (fig. 8.7). The length along the abscissa allowed for each stage is roughly proportional to the number of ammonite zones, this being thought to be a more reliable indicator of time values than treating each stage as equal. A fine grid was drawn on tracing paper and used to measure relative areas for given stages (see Hallam 1969b). The areal distribution of marine deposits for the other stages or substages was assessed with respect to these figures and the data plotted accordingly.

Fig. 8.7 brings out clearly a gradual transgression over the continents from the Hettangian to the late Oxfordian, only interrupted by a notable regression in the early Bathonian, and then a pronounced regression at the end of the period. Possibly the extent of the early Bathonian regression is overestimated, because new collecting and improved zonation has proved the presence of deposits of this age in areas where they were formerly thought to be absent, although this has been allowed for as far as possible. One cannot readily discern transgressions or regressions on a world-wide scale related to specific stages. Even the 'Callovian transgression' is seen to begin in the late Bathonian, its effect being enhanced by the preceding regression.

Recognition of the overall pattern is not new. It is apparent, for instance, in the palaeogeographic maps of Termier and Termier (1960), while the data of Ronov (1968) indicate that the sea attained successively

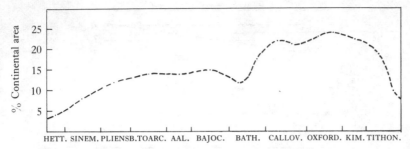

FIGURE 8.7 Graph to illustrate approximate percentage continental area covered by sea for successive Jurassic stages.

greater extent in the Lower, Middle and Upper Jurassic. The maximum extent of Jurassic seas was assessed by Ronov at 33 per cent of the total continental area, by the Termiers at 27 per cent, and is assessed here at 24 per cent. Ronov's figure is almost certainly an overestimate, and the present data may well be a slight underestimate, but we can accept with confidence that the Jurassic seas at their maximum extent covered between a quarter and a third of the total continental area.

The underlying control

A transgression or regression in a given area may result either from local epeirogenic warping or from a world-wide eustatic change. The presence of angular unconformities, difference of metamorphic grade or introduction of coarse clastic sediments into a sequence suggest local tectonic activity, but in the majority of cases in the tectonically relatively quiescent Jurassic such evidence is lacking, and transgressions and regressions can often be correlated more or less precisely by ammonites over extensive regions, if not intercontinentally. In such cases the possible operation of eustatic control must be taken into account.

Regressions are relatively difficult to assess, because orogeny is associated with regional uplift. Consider the models illustrated in fig. 8.8, which embody principles generally familiar to stratigraphers. A number of simplifying assumptions are made. Epeirogenic movement about a hinge position H causes uplift in the hinterland to be balanced by subsidence in the area of marine deposition, with sedimentation keeping pace with subsidence; clastic sediments are carried to the sea from the land. Of course, uplift and subsidence may not be thus related, sedimentation may not keep pace with subsidence and, if drainage from the hinterland is at a minimum, transgressive carbonates may come to lie directly on the old land surface. However, the diagrams illustrate how in many instances tectonic

173

Changes of sea level

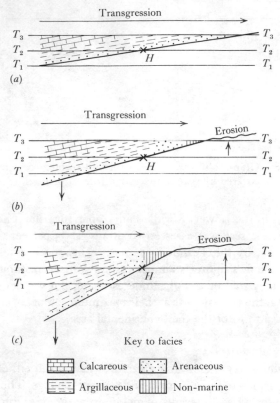

FIGURE 8.8 Diagrammatic erosion–sedimentation models to illustrate the counteracting effect of epeirogenic uplift in the hinterland of a steadily transgressing marine transgression: (*a*) no local epeirogeny; (*b*) minor local epeirogeny; (*c*) major local epeirogeny. *H*, hinge zone.

uplift will have the dual effect of modifying the sedimentary facies and counteracting the spread of sea consequent upon a gradual eustatic rise of sea level.

Turning to the actual situation in the Jurassic, one is struck by the close correlation in time between the pronounced regression near the end of the period and the most intense phase of orogenic activity, as recorded in the last chapter. Two interpretations can be suggested as follows. A phase of tectonic activity characterised by intense orogeny in certain geosynclinal regions was manifested over extensive sectors of more stable regions by gentler epeirogenic uplift, which would suggest tectonism on a world-wide scale. Alternatively, orogenic uplift in the geosynclines was manifested on the adjacent ocean floor by the subsidence of deep trenches, thereby provoking a eustatic fall of sea level. Quite possibly both factors have played a role. Similar explanations can be suggested for the more modest Middle Jurassic regression.

With regard to the dominant Jurassic motif of transgression, it appears more plausible to invoke a eustatic rise of sea level than a more or less uniform subsidence of the continents. Admittedly, regional subsidence and relief must have played a role in controlling the variable extent of transgression from one stage to the next in different parts of the world, but this appears to be only a subsidiary factor. From the discussion in the first part of this chapter it is evident that an intermittent eustatic rise combined with local subsidence and sedimentation is sufficient to account for many sedimentary cycles, without periodic lowering of sea level as well. We may next attempt to assess the approximate extent and rate of the Jurassic rise in sea level. It will be seen from fig. 8.7 that the sea spread from an area of less than 5 per cent of the total continental area at the beginning of the period to almost 25 per cent in the Upper Oxfordian. According to Kuenan's (1950) estimate, a rise of about 50 m would cause the sea to cover about 25 per cent of the continents in most periods of geological time. Accepting this estimate, and a length of about 60 million years for the period (Howarth 1964), it can be inferred that sea level might have risen almost 50 m in about 40 million years from the Hettangian to the Oxfordian, a mean rate of rise of almost 1.25 mm per 10^3 years.

Eustatic changes can result either from the melting and freezing of polar ice caps or from changes in the cubic capacity of the ocean basins. Bearing in mind the evidence reviewed in chapter 9 of Jurassic equability, and lack of evidence of ice caps, the latter explanation seems the only reasonable one (Hallam 1963b). With the general acceptance of plate tectonics, it becomes reasonable to relate sea level rise to uplift and lowering to the subsidence of ocean ridges (Hallam 1969b, Brookfield 1970).

With our present knowledge of the epeirogenic mobility of oceanic ridges and trenches there is no particular difficulty in invoking this factor to account either for laterally extensive sedimentary cycles or for transgressions and regressions, many of which, in a regime of shallow epeiric seas, would have required changes of sea level of no more than a few metres over a period of time ranging from several hundred thousand to a few million years. Originally, an attempt was made to relate the Jurassic rise in sea level to uplift of the Darwin Rise in the Pacific (Hallam 1969b) but, in the light of new information (reviewed in the last chapter) throwing doubt on its existence, this interpretation must be abandoned.

In the model proposed by Valentine and Moores (1972) eustatic rises of sea level should result from ocean ridge uplift at times when continents are moving apart, which is what they were beginning to do in the Jurassic. They estimated that the total volume of present ridges is equivalent to a change of sea level of approximately 74 m over the oceans or 52 m over the entire earth's surface. This would appear to be an underestimate, according to the recent work of Hays and Pitman (1973), who established a correlation between rates of sea floor spreading and epeirogenic movements of the ocean ridges. Acceleration of spreading rate in the late Cretaceous relates to uplift of the ridges and hence to a major eustatic transgression, while subsequent retardation of spreading rate has led to ridge subsidence and regression. Hays and Pitman computed the total volume of the present ridge system and the change in continental freeboard that should result from a given change in ocean depth. Fig. 3 of their paper is a graphical plot of the inferred changes in sea level since the early Cretaceous for given areas of present land surface covered by sea. From their data we can reckon that sea level rose about 150 m from the Hettangian to the Oxfordian, to reach a level over 200 m higher than the present. These figures are three times higher than Kuenan's, and would imply a rate of rise in sea level of nearly 4 mm per 10^3 years. As we have today much fuller information than was available to Kuenan, the Hays and Pitman data may provisionally be accepted as more accurate.

RELATIONSHIP WITH FAUNAL CHANGES

The changes of fauna from one part of a given sedimentary cycle to another, as documented in detail for instance by Sellwood (1970), are obviously an expression of changing environments from shallower to deeper water or vice versa. A correlation has also been noted between major cycles and the ammonite sequence (Hallam 1961, 1963b). Frequently the top of a given cycle corresponds with the termination of an ammonite zone or stage. Regional extinction often appears to be bound up, in fact, with the shallowing phase of a given cycle, a phenomenon which may be enhanced or sharpened, but not wholly created, by condensation or non-sequence at the cycle tops. Where the sedimentary change is pronounced other groups besides the ammonites may be affected. The change from the Pliens-bachian to the Toarcian in northwest Europe, already discussed earlier in the chapter, provides a well-documented example (Hallam 1961, 1967c, 1972a).

The sedimentary changes in question, from shallower to deeper water facies, correlate closely with a profound turnover of the invertebrate fauna, with very few species common to both the Pliensbachian and Middle Toarcian. Ammonites apart, the major faunal break occurs at the level of a widespread horizon of laminated bituminous shales (discussed in chapter

3). These are almost devoid of benthos and the question arises as to what extent the evidently hostile depositional environment was responsible for the widespread apparent extinction.

Some evidence exists in support of this idea. The Lower Toarcian bituminous shale facies does not extend southwards beyond southern France, and in northeastern Spain a number of abundant bivalve species, which disappear further north at the top of the Pliensbachian, persist into the Middle Toarcian (Bifrons Zone). On the other hand, a considerable number of new species of ammonites, brachiopods and bivalves enter the succession in the Bifrons Zone over a wide area of southwest Europe, and the bivalves persist into the Bajocian. The most striking change among the ammonites occurs at the base of the stage, with the northward spread of descendents of Tethyan hildoceratids and dactylioceratids following the extinction of the boreal amaltheids. Thus, although the contrast is less marked in southwest than in northwest Europe, the faunal turnover from the late Pliensbachian to the Toarcian is still considerable, and more marked than between many other Jurassic stages.

It is tempting to relate the widespread changes across the Pliensbachian–Toarcian boundary to the initiation of opening of the southern North Atlantic and tectonic collapse of the carbonate platform along the western margins of the Tethys. There are various indications, such as stratigraphic gaps and widespread shallow water ironstones and sandstones, to suggest that the Toarcian transgression due to rise of sea level was preceded by a regression, which might have been an important factor in the extinction of boreal ammonites and other faunal elements.

A very comparable facies change takes place from the late Oxfordian to the early Kimmeridgian in southern England, and Brookfield (1973*b*) likewise found that, in the Mutabilis Zone of the Kimmeridgian, a major faunal turnover coincided with an episode of widespread deposition of bituminous shales.

Changes of the type discussed, but on a larger scale, involving mass extinction followed by evolutionary radiation of the survivors, have been recognised in various parts of Phanerozoic time and also related to eustatic changes of sea level (Newell 1967; see also Valentine and Moores, 1972 and Hays and Pitman 1973). A fascinating indirect link between faunal extinctions and plate tectonics is thereby suggested, which is worthy of much more thorough and extensive investigation than has so far been attempted.

9   Palaeoclimates

There is widespread agreement that the Jurassic climate was appreciably more equable than today's, with no polar ice caps and hence a much more modest decline of air and sea temperatures with increasing latitude. In this chapter the evidence from fossils, sediments and oxygen isotope geochemistry will be critically assessed with a view to seeking confirmation and perhaps amplification of this general belief. In addition, more particular hypotheses will be considered, for instance that which attributes the paucity of limestones or corals in the European Lias or in the Jurassic of northern Europe and the Arctic to a cooler climate, and that which attributes the late Jurassic geographical spread of carbonates and evaporites at the expense of coals and ironstones to increasing aridity.

EVIDENCE FROM FOSSILS

Terrestrial plants

Since terrestrial floras include many extremely sensitive climatic indicators, and since Jurassic plants have a reasonable number of living relatives, they should provide perhaps the best indicators of Jurassic climate.

As is well known, the Jurassic was an age of gymnosperms, in the form of cycadophytes, gingkophytes and conifers, but ferns and horsetails were also a major component of the flora. The fact that rich floras are known both in Grahamland (63° S) and the New Siberian Islands (75° N) is in itself a strong argument in favour of a more equable climate than the present.

Among the ferns there are abundant widespread genera whose living relatives cannot tolerate frost. In particular, the Matoniaceae and Dipteridaceae are at present restricted to the tropical Indo-Malaysian region but were distributed over a much wider range of latitude in the Jurassic. In the Liassic it is possible to distinguish within the Northern Hemisphere a northern floral zone embracing Greenland, Sweden, central Europe, Siberia and Japan, and a southern zone extending from central America to the Middle East and southern China (Wesley 1973, Barnard 1973). This presumably reflects a degree of climatic differentiation, although Barnard (1973) considered that the difference in the two floras might relate in part to the continentality of the climate, i.e. to the degree of contrast between seasons.

Evidence for latitudinal climatic differences in the Northern Hemisphere comes also from the cycadophytes (mostly Bennettitales, true cycads being rare). These tend to be restricted, according to Vakhrameev (1964, 1965) and Wesley (1973), to a belt including Mexico, Europe, the southern USSR and Japan. Passing northwards into Siberia, the cycadophyte and conifer floras are much poorer in diversity and are instead dominated by gingkophytes. This signified to Vakhrameev a latitudinal climate gradient in Eurasia significantly less than today, with winter temperatures in Siberia probably never falling below 0 °C. Seasonality in this region is shown by annual rings preserved in coniferous wood. The climate of this northern zone appears to have been humid and moderately warm, while that of the zone to the south compares with the present humid tropical–subtropical zone.

With regard to the Rhaeto-Liassic of the Southern Hemisphere, the floras which contain *Dictyophyllum* differ considerably from those of the Northern Hemisphere, and appear to mark a separate floristic province. The wide spread of *Dictyophyllum* to between 50° and 60° on either side of the equator is a strong argument in favour of a warm and equable world climate (Barnard 1973).

As regards the conifers, Florin (1963) has conclusively established a differentiation between the Northern and Southern Hemispheres, the Taxodiaceae, Cupressaceae and Pinaceae being confined to the former and the Podocarpaceae to the latter, together with India. The Indian floras in general are indeed significantly different from those in adjacent Afghanistan and the USSR, which presumably relates to the subcontinent's northward migration across the Tethys in post-Jurassic times. The Indian Jurassic climate was evidently of tropical humid type, an interesting point bearing in mind that, apparently, the site of Indian plant-bearing deposits was in Jurassic times over 40° S. The contrast between the Northern and Southern Hemispheres evidenced by the conifers is unlikely to relate simply to climate, because climatic control should have given rise to a symmetrical latitudinal pattern of zones with respect to the equator. It is more likely to be due to the substantial, and ultimately complete, separation of Laurasia and Gondwanaland by the Tethys, and inability of pollen to cross the marine barrier.

Little evidence of Jurassic arid zones has so far been forthcoming from the plants, but Vakhrameev noted the appearance in the Upper Jurassic of the southern USSR of *Stachypteris* and *Lomatopteris*, with a postulated xerophytic habit. There was at this time a small northward shift of the southern boundary of the Siberian floral zone.

Palaeoclimates

Reptiles and fish

Rich dinosaur faunas are known only from the Upper Jurassic of the Western Interior of the USA (Morrison Formation) and Tanzania (Tendaguru Beds) but scattered records exist over a wide latitudinal range, from the USA, Europe and China to Patagonia and Australia. Charig (1973) was unable to recognise genuine provinciality and several families, and perhaps many more genera, were cosmopolitan. Of particular interest is the close similarity of the two faunas mentioned above, with genera such as *Brachiosauras* and *Barosauras* (and probably a number of others) common to the two. With such large terrestrial animals of low reproductive potential, this fact almost certainly implies a free land connection until quite late in the Jurassic, which poses a problem. With the Smith *et al.* (or any similar) reconstruction, adapted to allow a greater separation of Africa–South America away from North America, the only possible crossing point would seem to be between southwest Europe and North Africa. Unfortunately the Jurassic deposits in the critical region are largely marine. Presumably one has to postulate periodic emergence of land bridges from a shallow sea.

Dinosaurs are usually considered to have been ectothermic, relying mostly on external heat sources for their temperature regulation. As large reptiles are today confined to low latitudes because of their intolerance of cold, they make good climatic indicators (Romer 1961). Widespread latitudinal distribution of Jurassic dinosaur genera would therefore appear to support the notion of equability.

Recently, however, Bakker (1972) has argued on the basis of anatomy and ecology that the dinosaurs were in fact endothermic like mammals, using a high endogenous heat production to maintain a constant body temperature. They are likely, nevertheless, to have been stenothermal, because the combination of large size and naked skin would have rendered them unable to tolerate prolonged drops in their body temperature, such as would occur during the winter seasons of even cool temperate zones.

Schaeffer (1970) has discussed the evidence of Mesozoic fish remains and concluded that the extensive north–south distribution of ceratodontids and certain actinopterygian groups must signify an essentially uniform climate embracing all the continents. Considering for instance the Jurassic ceratondontid lung fishes, these were more or less world-wide in distribution, with records from China, England, the United States, Madagascar and Australia; their living relatives are confined to the tropics or subtropics.

Marine invertebrates

The bearing of temperature change on the distribution of marine invertebrates will be discussed at some length in the next chapter, and attention
will here be confined to corals and molluscs.

Hermatypic corals cannot survive at temperatures below about 16 °C and
significant reef development only takes place above 18 °C (Wells 1967).
Coral reefs are therefore confined today to the tropical–subtropical zone,
from about 30° N to 30° S. It is therefore of considerable interest that
Upper Jurassic reefs occur as far north as Sakhalin (Beauvais 1973), about
60° N on the Smith *et al.* reconstruction. The absence of corals from
comparably high latitudes elsewhere is likely to be due to factors other than
temperature, as will be discussed in chapter 10.

As Arkell (1956) observed, giant shelled molluscs, like hermatypic
corals, are also confined to the tropics and subtropics at the present day, so
that their much wider distribution in the Jurassic provides further support
for a more uniform climate. Detailed microstructural and mineralogical
research on well-preserved molluscan shells can also confirm this, as well
shown by the work of Hudson (1968) on aragonitic shells of the bivalve
Praemytilus strathairdensis from the Bathonian of Scotland. The fact that
the shells are wholly aragonitic suggests, by comparison with living *Mytilus*
and *Modiolus*, a subtropical or tropical environment. Temperate zone
varieties have a calcitic outer layer. The presence of fine growth banding
suggests seasonal temperature fluctuations.

EVIDENCE FROM SEDIMENTARY ROCKS

Besides the references on particular topics cited below, use has been made
of the compilation of data on Jurassic world lithofacies by Ronov and Khain
(1962). This work contains valuable data on the thickness and lateral
distribution of a wide variety of both terrestrial and marine rocks, though
one cannot vouch for its accuracy. For the Soviet Union, the comprehensive and sumptuously produced lithological–palaeogeographical Atlas
(Vereshchagin and Ronov 1968) proved of help for a part of the world which
is inadequately described in the English language. The distribution of
important deposits of climatically significant rock types is given in figs. 9.1
and 9.2 which utilise the Smith *et al.* reconstruction for the early Jurassic.
Data for the Lower and Middle Jurassic are combined, because the major
lithofacies distributions are closely similar, and the reconstruction can be
considered highly suitable since it is meant to represent the position of the
continents at about 170 million years ago. The Upper Jurassic map is less

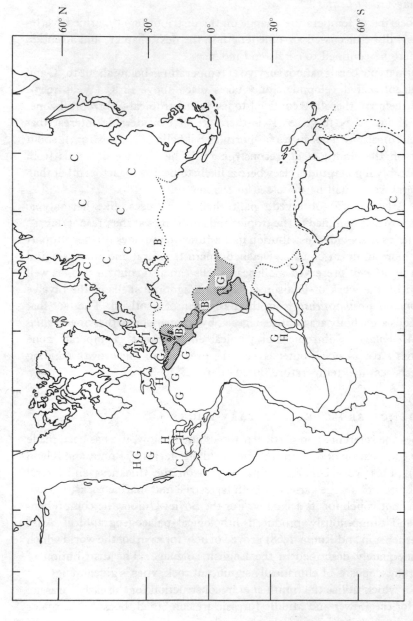

FIGURE 9.1 Distribution of principal deposits of climatically significant rocks in the Lower and Middle Jurassic. B, bauxite; C, coal; G, gypsum or anhydrite; H, halite; I, ironstone; stippled, principal zone of carbonate sedimentation.

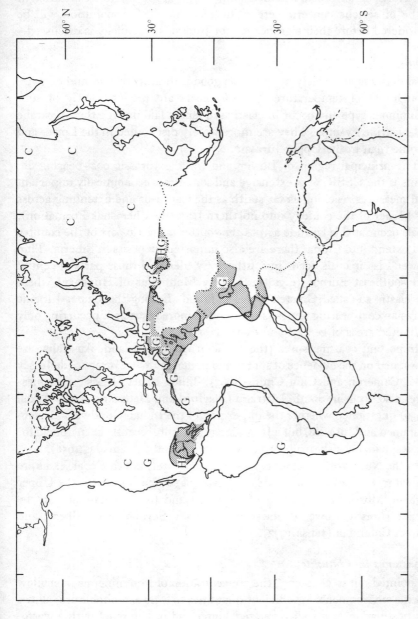

FIGURE 9.2   Distribution of principal deposits of climatically significant rocks in Upper Jurassic. Key as in fig. 9.1.

accurate, because the gap in the southern North Atlantic between Africa and North America had widened somewhat, but the difference is too slight to be of serious concern here. The main facts of distribution will be considered before their significance for Jurassic climatology is discussed.

Coals

Coal deposits are widely accepted as good indicators of a humid climate, though not of temperature. Jurassic coals are predominantly of sub-bituminous type and occur in clastic sequences laid down either in paralic or lacustrine swamps. They are more widely distributed in the Lower and Middle than in the Upper Jurassic.

The principal region of Lower and Middle Jurassic coal-bearing deposits is the USSR, with extensive and sometimes economically important coal measures reaching as far south as the Caucasus and extending across the border of the country into northern Iran (the Shemshak Formation). Coals occur at this latitude across the southern Asiatic part of the country and extend into China; there are also important deposits in Siberia. Thin Lower Liassic coals occur in southern Sweden and those parts of central and southeast Europe (e.g. the Villany Mountains of Hungary) where the clastic so-called Gresten facies is found. During the Upper Jurassic Eurasian coal-bearing deposits were much more restricted, occurring only in limited areas of central and eastern Asia.

Important coal measures (the Walloon) in Queensland, Australia, are now dated on the evidence of spores and pollen as Lower Jurassic (de Jersey 1960, Cameron and Chiu Chung 1963). Thick Rhaeto-Liassic coal measures also occur in South Australia (Playford and Dettmann 1965). Arkell (1956) questioned whether the upper part of the Indian Gondwana System contained any Jurassic, but a Jurassic age has more recently been upheld for the Rajmahal coal-bearing beds (Mahadevan and Srivamadas 1958).

In the New World coal measures are much rarer, but thin coal seams are known in the Lower and Middle Jurassic of southern Mexico, the Upper Jurassic Morrison Formation of Montana and the Dakotas, and in the Upper Jurassic–Lower Cretaceous Kootenay Formation of Alberta and British Columbia (Jansa 1972).

Ironstones and bauxites

As pointed out in chapter 3, the concentration of iron minerals in shallow marine environments in sufficient quantities to form exploitable ores allows the postulation of a well-vegetated hinterland of low relief with a warm, humid climate. The abundance of kaolinite in the clays associated with the European ironstones is consistent with this interpretation. At the present day it occurs in greatest concentrations on the sea bed near the coasts of tropical and subtropical lands, where it has been produced in the soil as a result of intensive chemical weathering in an acid environment.

The distribution of sedimentary iron ores in northwest Europe has already been outlined in chapter 3. The iron minerals chamosite and siderite are associated with sandy deposits throughout a much larger area in Eurasia, extending as far south as the external Alpine–Carpathian zones, the Balkan Mountains and the Caucasus. The ironstone facies is clearly closely related in space and time to the coal measures facies (which usually contains abundant siderite in the form of concretions). Thus the areal extent and volume of ironstones diminishes from the Lower and Middle into the Upper Jurassic.

Lower Jurassic bauxites occur in southern Uzbekistan and were considered by Ronov and Khain (1962) as further indicators of a humid climate. They are also recorded from a number of Jurassic horizons in Yugoslavia in association with marine platform carbonates (Grubić 1964). The same is true of Greece, Turkey, Iraq and Israel; in the last-named country the bauxites occur in the form of vadose pisolites (Goldberg and Friedman in press). The bauxites presumably signify intervals of emergence and intensive chemical weathering in a warm climate with a humid season.

Evaporites

Substantial deposits of evaporites provide the best indicator of warm, arid climates. At the present day two such zones exist flanking the equatorial humid zone. In the Northern Hemisphere the southern limit of the arid zone varies between about 15° and 20° N and the northern limit from about 35° N in the Mediterranean region to 40° N in western North America and central Asia. In the southern Hemisphere the northern limit lies similarly at about 15° to 20° S while the southern limit reaches as high as 50° S in western South America. Lotze (1964) has demonstrated that a good correlation exists between the two warm arid zones and the distribution of Quaternary evaporites.

Turning to the Jurassic evaporites, we shall consider firstly the American continents. In the Western Interior of the USA important gypsum deposits of Bajocian age occur over a wide area in Utah, Wyoming and South Dakota. Gypsum and anhydrite occur also in the Callovian and Oxfordian. Thick deposits of the latter stage are known in New Mexico, and thinner deposits occur also in Utah, Arizona and Colorado. Substantial halite deposition took place in Callovian times in Utah and adjacent parts of Idaho and Wyoming.

Palaeoclimates

A thick halite and $CaSO_4$ sequence occurs also around the margins of the Gulf of Mexico. In Louisiana it is known as the Louann Salt and is tentatively dated on the basis of spores as late Triassic to Middle Jurassic (Lehner 1969). The first marine horizon above the salt is Upper Oxfordian. The Todos Santos Formation of southeastern Mexico has been shown by drilling to contain substantial salt deposits intercalated in red beds. The formation is dated as post-Callovian, pre-Kimmeridgian, by Viniegra (1971). The Todos Santos salt could therefore be correlative with the Louann salt, or slightly younger. Salt-bearing red beds in Cuba are considered to be probably post-Rhaetian, pre-Callovian in age (Khudoley and Meyerhoff 1971). Recent drillings on the Nova Scotia Shelf have revealed over 750 m of more or less pure halite, named the Argo Salt, which is overlain by an anhydrite–gypsum and carbonate sequence. The salt is not well dated but is believed on palynological evidence to be Lower Liassic and probably pre-Pliensbachian in age (McIver 1972).

In South America thick gypsum and anhydrite deposits are found in northern Chile and are dated as Upper Oxfordian to Kimmeridgian (Arkell 1956). Other Jurassic red beds with gypsum intercalations are known from elsewhere in the Andean region (e.g. Geyer 1973).

As regards Europe, thin beds of gypsum and anhydrite associated with dolomite and red beds range up into the Hettangian in Portugal and have been found in boreholes in the Aquitaine Basin and the Pays de Bray, in the Paris Basin, where they are also dated as Hettangian (Bouroullec and Deloffre 1969, Lefavrais-Raymond and Horon 1961). In the top Jurassic Purbeck beds of southern England anhydrite has been discovered in boreholes and pseudomorphs after gypsum and halite are quite common in surface outcrops. The gypsum-bearing Oncala Beds of the Iberian Ranges of northeast Spain are roughly correlative, being dated as Tithonian on the basis of ostracods (Kneuper-Haack 1966). In northwest Germany halite and gypsum occur in the Münder Mergel, which is thought to correlate with the Lower Purbeck Beds of England (Arkell 1956), and anhydrite is recorded from boreholes in the Kimmeridgian and Tithonian of Aquitaine (Bouroullec and Deloffre 1969).

Important evaporite deposits also occur in the southern USSR. $CaSO_4$ appears for the first time in the Oxfordian and Kimmeridgian in a zone extending west from the Caucasus into Moldavia and east into Uzbekistan. Anhydrite and gypsum continued to be deposited in Tithonian times in the Caucasus, with only minor quantities of halite, while a major salt basin developed further east in Turkmenia and southern Uzbekistan. Subordinate qualities of potash salts occur in association with thick halite–anhydrite deposits.

186

The other part of the world where drilling in recent years has revealed the existence of significant evaporite deposits is around the margins of Africa. Over 2500 m of pre-Bathonian gypsum and halite-bearing beds, probably ranging down into the Triassic, occur under southern coastal Tanzania (Kent *et al.* 1971). Similarly, in the Tunisian Sahara, a gypsiferous sequence 600 m thick and resting on Triassic beds is overlain by marine Bathonian carbonate which themselves contain some gypsum (Busson 1967). Lower and Upper Jurassic gypsum and anhydrite also occur locally in the zone extending from western Morocco to Senegal (Reyre 1966).

Finally, an important evaporite formation, the Hith Anhydrite, has been found by drilling in southernmost Iran, and has been dated as Jurassic (James and Wynd 1965).

Aeolian sandstones

The only region where reasonably well-authenticated Jurassic aeolian sandstones are known is the US Western Interior (see chapter 6). The measurements on dune bedding orientations are broadly consistent with the region lying in a trade wind belt at that time. Bigarella and Salamuni (1964) determined palaeo-wind directions for the apparently aeolian and presumed Triassic–Jurassic Botucatu Sandstone of the Paraná Basin of South America. This formation is associated, however, with the Serra Geral basalts, which are now known to be Cretaceous in age.

Limestones

Bearing in mind the distribution of Recent carbonate deposits, the presence of thick limestone sequences has generally been held to signify a warm climate, and their absence a cool climate. While the former interpretation is reasonable enough the latter is highly questionable as a general inference, because the deposition of carbonate as opposed to terrigenous clastic sediments is controlled in the first instance by what might be termed the tectonic–sedimentational regime. Thus while the southern side of the Persian Gulf is an area of carbonate deposition the northern side is not, being influenced by the Tigris–Euphrates delta system; the climate is nevertheless arid throughout. Likewise, deposition of carbonates around southern Florida and the Yucatan gives way progressively towards the Mississippi delta to sand, silt and clay deposition, within the same broad zone of humid climate. Furthermore, the circum-Pacific eugeosynclines are, in both high and low latitudes, zones of volcaniclastic and quartz clastic sediments.

The principal zone of Lower and Middle Jurassic carbonate deposition was in the circum-Mediterranean region extending eastwards into Saudi

Palaeoclimates

Arabia and southern Iran, that is, according to the continental reconstruction adopted in this book, around the western and southern borders of the western Tethys.

Upper Jurassic times saw the expansion of this zone westward into Central America and northward into the southern part of Eurasia, with concomitant retreat of the terrigenous clastic facies; the change from the 'Black' and 'Brown' to the 'White' Jurassic of the old German classification is more than a local phenomenon.

Phosphates

The concentration of phosphorite ('collophane') nodules is primarily a function of slow sedimentation in a moderately shallow marine regime. Since modern phosphorite concentrations appear to be related to zones of oceanic upwelling related to divergence, such as off the western coasts of the trade wind belts, Ronov and Khain (1962) have concluded that an increase in the concentration of phosphorite nodules towards the top of the Jurassic, observed for instance in the condensed Volgian deposits of the Russian platform, signifies a change to a more arid climate.

Tillites

Unlike in many other geological systems, no well-authenticated tillites have been recognised in the Jurassic. An earlier claim that a rudaceous deposit in Victoria Land, East Antarctica, was glacial in origin (it was hence termed the Mawson Tillite) has been convincingly refuted by Borns and Hall (1969) and Ballance and Watters (1971). It appears in reality to be a volcanic deposit of Lower or Middle Jurassic age, probably of explosion breccias emplaced by gravity flow following extrusion.

Likewise, Cecioni's (1958) claim that late Jurassic tillites occur in Patagonia must be treated with scepticism. The critical evidence of striated pavements is lacking and the associated facies is of flysch type, which suggests a depositional environment of tectonic instability, with the deposits in question probably being emplaced by mudflows.

Discussion

Since it is quite evident from the fossil data that a warm climate extended over a much wider range of latitude in the Jurassic than today the sparsity or absence of limestones in Jurassic high latitude belts must be attributed to factors other than low temperature, such as those mentioned earlier. Likewise the relative geographic restriction of calcareous facies to the Tethyan belt in the Lower Jurassic need not necessarily signify that the climate was cooler at this time.

The spread during the Jurassic of calcareous at the expense of terrigenous clastic and ferruginous facies, which became pronounced in the Oxfordian in the southern part of Eurasia, is more plausibly explained by the secular rise in sea level inferred in chapter 8. As the epicontinental seas spread more widely, so the sources for sand, clay and iron minerals correspondingly diminished in area.

This interpretation also has a bearing on the inference by Russian workers (Ronov and Khain 1962, Strakhov 1967) that the late Jurassic climate was more arid than earlier in the period. Such an inference depends primarily on a correct evaluation of the climatic significance of the evaporites.

It will be seen from figs. 9.1 and 9.2 that the Jurassic evaporites generally lie within a range of latitude similar to that defined by the northern and southern limits of the warm arid zones today. Unlike today's pattern, however, in which the arid zones are separated by an equatorial humid zone, the belt of Jurassic evaporites extends around the western margin of the Tethys and appears to cross the equator, from Tunisia to Iran and Tanzania.

Although, as Ronov and Khain indicate, the proportion of evaporites (and carbonates) increases slightly in the Upper as compared with the Lower and Middle Jurassic, the latitudinal range is hardly greater. Furthermore, there are reasons for questioning whether the presence of evaporite deposits should necessarily lead to the inference of aridity on a large regional scale. Consider, for instance, the Hettangian anhydrite discovered in northwestern France, which actually lies to the north of correlative coal measure facies in southeastern Europe. Similarly, the lower Kimmeridgian Abbotsbury Ironstone and the evaporite-bearing Purbeck Beds of Dorset were only separated in time by a few million years. Are we to infer a major climatic change in southern England only on the basis of this? The detailed study of deposits laid down in environments marginal to the open sea, such as the Purbeck Beds and Great Estuarine Series, reveals a complex pattern of environments varying from brackish or fresh water to hypersaline (see chapter 4). As regards the New World, it is true that in southern Mexico Lower and Middle Jurassic coal-bearing deposits are followed by Upper Jurassic saliferous beds, but the reverse is the case in the northern part of the US Western Interior.

In southern Israel the sedimentary sequence passes upwards from early Jurassic anhydrite and stromatolite-bearing carbonates to middle or late Jurassic sandstones with a few thin coal seams, indicating a change in environment from sabkhas to a fluvial and salt marsh regime (Goldberg and Friedman in press). In the Goldberg and Friedman paper a comparison is made with the modern Gulf Coast region. In the Texan Laguna Madre the climate is arid and gypsum is being deposited, while in Louisiana the climate is humid and salt marshes are abundant; hence the upward change in the Jurassic succession in Israel is held to signify a climatic change from arid to humid. This may well be valid on a local scale, as indeed it might be for the reverse situation in southern England, but it must be remembered that the influx of freshwater from a large river entering an arid region will inhibit evaporite and carbonate deposition and favour the local development of short-lived swamps. The inference of regionally extensive climatic belts can only be made confidently with the evidence of thick and widespread sediments of the type generally held to be good climatic indicators.

Quite generally, evaporites are found in environments onshore from carbonate platforms or banks, and coal measures are the paralic equivalent in regimes of terrigenous clastic sedimentation. The late Jurassic spread in areal extent of evaporites in southern Eurasia appears to correlate closely with the spread of limestones. The presence of evaporites in fact may be at least as much a function of local sedimentary environment as of regional climate. Shaw (1964) has indeed argued cogently that within a warm shallow epeiric sea with no major influx of fresh water from the land there should be a general tendency towards hypersalinity which would in particular circumstances favour precipitation of the less soluble salts without the climate necessarily being especially arid.

Ronov and Khain's subordinate argument favouring an increase in aridity in the late Jurassic, based upon phosphorites, is also open to question. It is certainly true in the British area that phosphatic nodules and phosphatised fossils are much more evident in the Volgian and Cretaceous marine beds than earlier, as is the glauconite with which it is usually associated. Casey (1971) preferred to interpret this, however, as the result of a change of oceanic circulation consequent upon the opening of the North Atlantic, and the upwelling of cool waters from the deep ocean on the western side of the Eurasian continent.

While there seems at present to be little good reason to believe in significant temporal changes in the Jurassic climate, an important regional change is discernible. Coal measures are largely confined to the eastern parts of Laurasia and Gondwanaland, evaporites largely to the west. The best-authenticated arid environments are in the US Western Interior, where both aeolian sandstones and thick evaporites, including halite, occur. Thick deposits of halite such as occur in Utah, around the Gulf of Mexico and in Uzbekistan, were probably laid down subaqueously in supratidal or interior depressions and must be considered good indicators of warm, arid conditions. The same is presumably true of the thicker $CaSO_4$ deposits.

We are led to infer the existence of a western arid belt and two eastern humid belts, as did Robinson (1971) for the Triassic on the basis of a similar distribution of coals and evaporites. It is therefore quite relevant to apply her interpretation of the world climatic pattern to the Jurassic.

Bearing in mind the climatic role of the Eurasian landmass today in creating monsoonal conditions to the south, Robinson considered that the two supercontinents Laurasia and Gondwanaland might have acted likewise, so that winds reaching the eastern parts in middle and low latitudes might have brought monsoon-type summer rains, whilst a dry hot season would occur in winter as winds blew offshore. The central and western parts of the supercontinents would have tended to have an appreciably less humid climate because many of the dominant easterly winds would have travelled over land for a considerable distance, or, blowing equatorward without the intervention of mountains, could not readily have jettisoned their moisture. Coals formed in the eastern, peninsular parts of the landmass in middle to high latitudes where the temperature was more moderate and the rainfall less strictly seasonal, being under the influence of westerly and polar easterly winds.

One interesting implication of this interpretation is that northern Iran, with its coal-bearing Shemshak Formation, lies more appropriately adjacent to the Laurasian than the Gondwana continent, thus supporting a similar inference based on faunal distribution (see chapter 7). Barnard (1973) commented on the northerly affinities of the Shemshak plants of the Kerman region, and argued that this ruled out a separation of this region from Eurasia by the Tethys. In fact, the region in question, though in southern Iran, lies north of the Zagros Crush Zone and is geologically part of northern Iran.

EVIDENCE FROM OXYGEN ISOTOPES

After an early period of enthusiasm about the prospects of determining palaeotemperatures from oxygen isotope ratios, disillusionment has set in at least as regards the Mesozoic, because of the various factors that can complicate interpretation and give spurious results.

Three basic assumptions have to be made before the $\delta^{18}O$ value, determined from fossil shells by mass spectrometry, can be validly translated into ambient water temperatures. The isotopic composition of sea water is assumed to have been the same as the present. This cannot be proven but is likely to be true to a first approximation. At least the investigator can ensure that his fossil samples come from a normal marine environment, as established by the presence of stenohaline organisms. Influx of fresh water from the land normally lowers the $\delta^{18}O$ values, because of an isotope fractionation effect as sea water is evaporated and eventually precipitated as rain on the land. This would give a spuriously high temperature determination. It has been estimated that a 5 ‰ reduction in salinity during non-glacial periods is equivalent to a temperature change of about 5 °C in terms of changing the ratio of ^{18}O to ^{16}O (Epstein and Mayeda 1953).

The second assumption is that the organisms have secreted $CaCO_3$ in isotopic equilibrium with sea water. Early investigations showed that this is manifestly not the case for certain groups such as corals and echinoderms, but physiological fractionation effects seemed to be insignificant for the molluscs, which in consequence have provided the principal material for analysis. The third assumption is that isotopic exchange subsequent to precipitation of the $CaCO_3$ is negligible. Post-depositional isotope interchange with ground waters will normally act to give spuriously high temperature values, because, compared with sea water, ground waters are normally relatively deficient in ^{18}O. This is the most serious difficulty of the three and is not readily allowed for because isotope interchange can take place without manifesting itself by visible recrystallisation.

The early work on Jurassic fossils, summarised by Bowen (1966), dealt exclusively with the calcitic guards of belemnites, as these were the remains of stenohaline molluscs which were thought to have suffered little post-depositional alteration. Even at the time it was evident that this work left something to be desired, because rather bold conclusions were often drawn on the basis of single determinations on isolated museum specimens from widely scattered parts of the world. Bowen's (1961) paper on the contribution of belemnite oxygen isotope studies to an improved understanding of Jurassic palaeoclimates illustrates this well.

Bowen noted that the average temperature range from the tropics to the poles at the present day is about 50 °C, which is considerably greater than the range of 17 °C determined from Jurassic belemnites from all over the world. Curiously, Bowen failed to point out that he was here comparing air temperatures with water temperatures. Again, a solitary belemnite from the Callovian of Alaska gave a value of 16.9 °C, which is lower than the European results (ranging mostly between 20 and 30 °C). This single result was held to provide support for cooler conditions closer to the poles. Yet the lowest value of all, 15.9 °C, came from a New Guinea specimen! This was not explained, neither was the fact that the minimum temperature value for Western Australia, several degrees nearer the South Pole, was 18.5 °C.

Although a number of his determinations exceeded 30 °C, which is an approximate upper limit of any reasonable temperature model, there was no evaluation of the significance of such results.

That Bowen disregarded post-depositional complications is evident from his attempt to record average temperature values from his range of readings for different areas and horizons. If diagenetic or epigenetic changes are likely to give spuriously high values then such averages are meaningless. This point was recognised in the thorough study of New Zealand belemnites undertaken by Stevens and Clayton (1971). Their apparent temperatures ranged from less than 15 °C to 37 °C, and they frequently found it impossible to predict from visual inspection which material was the most altered. Their solution was to accept the minimum temperatures determined as the only valid data.

By a very full review of the literature Stevens and Clayton reinterpreted the previous Jurassic European palaeotemperature determinations in terms of minimum values, which are treated as true values (fig. 9.3). Possible trends which suggest themselves from these revised data are temperature minima in the Hettangian–Sinemurian and Upper Bajocian (Russia, Germany) and Callovian–Oxfordian (Russia only), with maxima in the Toarcian–Aalenian (Germany) and Kimmeridgian–Tithonian (Russia). The fact that the quoted Russian and German results disagree appreciably is not encouraging. The revised data, furthermore, contradict Bowen's (1966) claim of a temperature decline from the Oxfordian to the Kimmeridgian.

The difference in interpretation that can arise from taking average rather than minimum temperature determinations as valid is brought out clearly by the work of Fabricius *et al.* (1970) on various calcitic fossils and carbonate sediments of the Alpine Upper Triassic and Lower Jurassic. By considering only average values, the authors detected a steady decline in temperature from about 25 °C in the Norian Dachsteinkalk to 18 °C in the Lower Liassic, with a subsequent rise in the late Pliensbachian. This interpretation agrees conveniently with Fabricius' (1966) claim that the cessation of reef coral growth at the close of the Triassic was due to cooling, although, as we have seen in chapter 5, there is a more plausible explanation for this in terms of tectonic subsidence.

If one now takes minimum rather than average values, the Norian is lowest (16.5 °C), the Upper Pliensbachian highest (20.3 °C) and the Hettangian–Sinemurian intermediate (19.2–17.5 °C). These revised results contradict not only the Fabricius story but also the results for southern Germany illustrated in fig. 9.3.

A further difficulty arose from Longinelli's (1969) investigation of

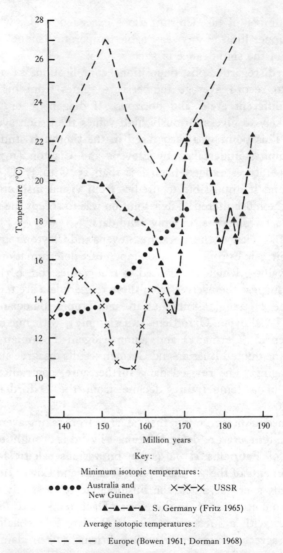

FIGURE 9.3 Compilation of oxygen isotope and average analyses of belemnite guards, giving minimum isotopic temperatures. (Adapted from Stevens and Clayton 1971.)

variations in $\delta^{18}O$ in the successive growth layers of belemnite guards. It had been widely accepted by earlier workers that regular fluctuations in $\delta^{18}O$ values signified cyclic temperature variations of several degrees centigrade, which were interpretable either as seasonal changes at the same site or seasonal migrations of the belemnite animals from warmer to cooler waters. Furthermore such regular cyclicity was thought to be a good indication that post-depositional disturbances had been minimal. Longinelli, however, found cyclicity with growth at 'temperatures' of well over 30 and even 40 °C. Since these results were so evidently spurious in terms of the original ambient water temperatures it was concluded that cyclicity in $\delta^{18}O$ values is in itself no criterion of lack of post-depositional alteration.

A recommendation that the problem of post-depositional alteration could only be solved by analysing aragonitic shell material or the calcitic shells associated with it in the same deposits was made by Jordan and Stahl (1969, 1970). They undertook detailed multiple analysis of several Jurassic ammonites from Germany and Poland which had survived the usual post-depositional fate of solution or recrystallisation to calcite. Their 'temperature' determinations for Pliensbachian ammonites ranged from 14.4 to 37.8 °C, with the higher values associated with a higher degree of calcitisation and hence rejectable as spurious. Of particular interest was a Callovian *Quenstedtoceras* from Poland, which showed only slight calcitisation; numerous determinations of material from both septa and outer wall gave a 'temperature' range of 7.7 to 17.1 °C.

The low temperature values were interpreted as signifying a cool climate associated with the Callovian 'boreal transgression', but low values were also obtained for Bajocian oysters (12.2 °C, 15.5 °C) and *Gresslya* (11.9 °C, 11.2 °C). In the context of what we can assume on other grounds about the European Middle Jurassic climate these figures seem anomalously low, and raise the question of whether there has not indeed been an appreciable physiological fractionation effect. Such an effect has been recognized between ammonites and *Inoceramus* in well-preserved aragonitic material in the Cretaceous of the US Western Interior (Tourtelot and Rye 1969). An investigation of comparable scope and thoroughness for the European Jurassic would undermine further our confidence in the validity of isolated data from belemnites alone.

However, Tan *et al.* (1970) were able to obtain climatically reasonable values of 20 to 25 °C for belemnites associated with aragonitic ammonites in the Callovian of Skye (but the ammonites gave temperature values 5 to 9° higher).

By far the most thorough analysis yet undertaken on the oxygen (and stable carbon) isotope composition of belemnites is that of Spaeth *et al.* (1971). These workers obtained numerous results from the different shell parts of two German Bajocian *Megateuthis*, both the aragonitic phragmocone and calcitic guard. Significant variation, of up to 5 ‰ $\delta^{18}O$ and 14 ‰ $\delta^{13}C$, was found to occur between guard and phragmocone. Cyclic variations were found to occur in the growth layers, but the guard (or rostrum) was argued to be almost certainly composed largely of secondary calcite, the original shell material during life being restricted to a spongy lattice work of crystals. Whereas the calcite gave apparent temperatures ranging from 9 to 25 °C, the aragonite values within the same specimen ranged even more widely from 14 to 43 °C!

This work, if it can be confirmed, strikes a severe blow at our already battered hopes for oxygen isotope geothermometry in the Jurassic. The more thorough the work, the more complications have arisen. While, taken in conjunction with stable carbon isotope analysis, oxygen isotope analysis may throw revealing light either on biochemical fractionation between or within organisms, or on the salinity regimes of marginal seas and lagoons (e.g. Tan and Hudson 1974), it seems extremely unlikely that it can add anything to what we can already infer from fossils and sediments about Jurassic climates.

CONCLUDING REMARKS

The evidence reviewed earlier in this chapter confirms conclusively that the general assumption of warmer, more equable conditions in the Jurassic as compared to the present day is fully justified. The tropical and subtropical zones were wider than today and temperate conditions, with no ice caps, characterised the polar regions. The zones of transition from easterly to westerly winds would have occurred several degrees nearer the poles than today. Schwarzbach (1963) reckoned that the decrease of mean surface air temperature from the tropics to the polar regions on an ice-free globe would be about 22 °C, compared with 42 °C at present. Such decrease in the meridional temperature gradient should lead to a decrease in intensity of the zonal winds. Because the ocean surface current system is driven primarily by these winds the currents should have been correspondingly more sluggish.

To conclude the chapter we shall enquire briefly into the likely pattern of the Jurassic ocean current system.

Luyendyk *et al.* (1972) have attempted an experimental simulation of the ice-free Northern Hemisphere circulation pattern for the Middle Cretaceous, using a planetary vortex model. As the configuration of the continents at this time did not differ drastically from that of the late Jurassic the more general results may be applied to the earlier period with reasonable confidence.

With a continuous equatorial oceanic belt separating Laurasia and Gondwanaland a zonal east to west Tethyan current is the dominant feature of the circulation pattern up to 20° N. North of this, both in the

Pacific and North Atlantic, clockwise-rotating gyres developed. In the case of the Atlantic, still at this time closed to the north, the Gulf Stream flowed as far as Newfoundland before turning east.

At the start of the Jurassic, of course, no world-encircling Tethyan current could have existed. Before the Atlantic began to open westward-flowing 'equatorial' water was presumably deflected both north and south by the land at the western end of the Tethys, to flow back eastwards as surface counter currents closer to the ocean coasts. Because of the much feebler contrast in temperature between equatorial and polar waters it is unlikely that the transport of water by surface currents from one latitudinal zone to another had much effect either on faunal distribution or regional climate.

10 Marine invertebrate biogeography

Since the pioneer study of Neumayr (1883) it has been widely recognised that many Jurassic marine invertebrate faunas are not cosmopolitan in distribution. Understanding the causes of provinciality remains one of the most intriguing problems demanding solution.

The older literature on provinciality is well summarised by Arkell (1956), Imlay (1965) and Stevens (1967). In this book we shall follow Arkell in using the term *realm* for the major divisions and *province* for the subdivisions of these realms. On the basis of ammonite distributions Arkell distinguished three realms, the Boreal, Tethyan and Pacific. Since the Pacific Realm is based on only a few ammonites in a fauna of overwhelmingly Tethyan affinities it is better relegated to at most a province of the Tethyan Realm (Hallam 1969a, 1971b). The Boreal Realm is well defined by ammonites and belemnites, but for other groups it is marked more by an absence of specifically Tethyan faunas than by distinctive genera of its own. It should also be noted that some ammonite groups were relatively cosmopolitan.

Since the subject is highly complex we must distrust facile explanations and reconcile ourselves to the limitations of the available data. Fossil collections are heavily weighted in favour of Europe and even in this continent important new discoveries and re-evaluations of data are still being made. Our knowledge of the taxonomy of many fossil groups is very inadequate, and even when a large taxonomic literature exists there is ample scope for disagreement in a field where subjective assessments still hold sway. For the most part, moreover, data on size, relative abundance and the associated bio- and lithofacies are sparse or absent.

Most of the information available concerns ammonites, due principally to their stratigraphic importance which has attracted a large number of specialists, not many of whom have evidently had a fine appreciation of the wide range of morphological variability shown by living species which are known to form discrete genetic entities, nor of the possibility of sexual dimorphism. One consequence of this is that the genus is the lowest taxonomic level that can be used reliably in intercontinental comparisons.

Even genera may be unreliable because some have been created on minute amounts of material by palaeontologists who lacked a comprehensive knowledge of the literature or of museum collections; very often the precise stratigraphic location has also been uncertain. Accepting, however, that most of the described taxa are reliable, interpretations must nevertheless be treated with even more than the usual caution, especially as critical environmental factors may not always leave their mark in the stratigraphic record. At the very least, such interpretations may provide a stimulus for the large amount of research still required.

THE BOREAL REALM

The Boreal Realm occupied the northern part of the Northern Hemisphere and is best defined by the distribution of the following ammonite families or subfamilies:

VOLGIAN	Craspeditidae
	Virgatitidae
	Dorsoplanitidae
KIMMERIDGIAN	Aulacostephaninae
OXFORDIAN	Cardioceratinae
CALLOVIAN	Kosmoceratidae
	Arctocephalitinae
BATHONIAN	Arctocephalitinae
BAJOCIAN (UPPER)	Arctocephalitinae
PLIENSBACHIAN	Amaltheidae
	Liparoceratidae

In a general way, the southern limit of these families corresponds in Europe approximately with a zone through southern Europe, and in the North Pacific region with a zone through northern California and between Japan and eastern Siberia. The boundary is, however, a gradational one and consequently there can be no general agreement on its precise location at a given time. Moreover, the boundary fluctuated in the course of time (fig. 10.1). Thus in the Callovian (especially the Upper Callovian) and Lower Oxfordian boreal ammonite genera reached far into southern Europe (the *boreal spread* of Arkell), while subsequently in the Upper Oxfordian it retreated back into northern Europe (this is more of a boreal retreat than a Tethyan spread, as Arkell thought; see Cariou 1973). Corresponding migrations are recognisable in North America. The situation is probably more complicated than that outlined by Arkell as the area of overlap with the Tethyan Realm also fluctuated on occasions.

Upper Bajocian and Bathonian boreal ammonites, not known when

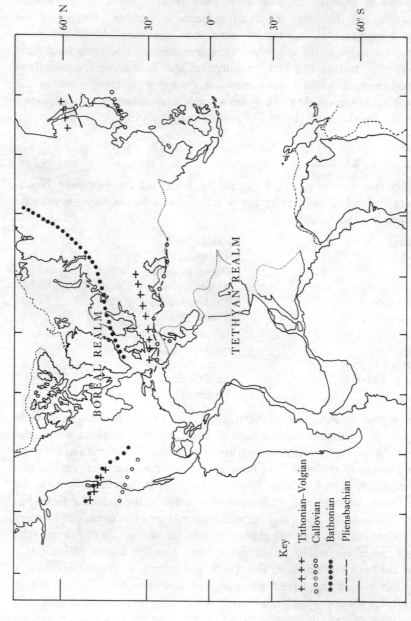

FIGURE 10.1 The approximate boundaries of the Tethyan and Boreal realms at different times in the Jurassic.

Key

++++ Tithonian–Volgian

ooooooo Callovian

●●●●●●● Bathonian

– – – Pliensbachian

BOREAL REALM

TETHYAN REALM

60° N

30°

0°

30°

60° S

Arkell wrote his review, are confined mostly to the Arctic regions. A distinctive boreal ammonite fauna first became clearly established in Pliensbachian times but was temporarily lost because of extinction of the Amaltheidae (Howarth 1973). Distinctness was renewed in the Middle Jurassic, reaching a maximum in the Bathonian, subsequently to become blurred somewhat in the Middle and Upper Oxfordian, and then set in once more to become extreme at the close of the period, in the Tithonian and Volgian. New elements were periodically recruited from the Tethyan Realm; thus the Aulacostephaninae and Dorsoplanitidae are two groups with Tethyan ancestors.

With regard to other Mollusca, belemnites show a clear faunal differentiation from Bathonian–Callovian times onwards, as marked by the Cylindroteuthidae (Stevens 1965, 1973). A notable southward migration of this family took place early in the Callovian, corresponding to the time of the boreal spread of the ammonites.

Specifically boreal elements apart from the molluscs are much less well documented on the whole, but Ager (1967) has pointed out that in the Pliensbachian of Europe such brachiopod genera as *Tetrarhynchia*, *Gibbirhynchia*, *Lobothyris*, indentate *Zeilleria* and ribbed *Spiriferina* have a predominantly boreal, or at least extra-Mediterranean, distribution, in so far as they are rare or absent in Ager's two inner circum- and intra-Mediterranean belts, though elements of the fauna do occur in Spain and Morocco. As regards the Foraminifera, Europe north of the Tethys is characterised by nodosariids and many simple-structured arenaceous genera, with periodic influxes of genera such as *Epistomina* and *Ophthalmidium* from Middle Jurassic times onwards (Gordon 1970).

There are indications of north–south provinciality in the ammonite faunas. Cariou (1973) distinguished a *Boreal Province* in the Callovian and Oxfordian in the Arctic, characterised by the dominance of Cardioceratidae, from a *Sub-boreal Province*, where kosmoceratids dominate in the Callovian and abundant perisphinctids occur in the Oxfordian. In a similar way Zeiss (1968) recognised, for faunas of late Kimmeridgian (or early Volgian) age, a Boreal Province embracing northern Europe, Greenland and northeastern Siberia, with such genera as *Pavlovia*, *Dorsoplanites*, *Laugeites* and *Subdichotomoceras*, and a Sub-boreal Province embracing western and eastern Europe, with genera including *Pectinatites*, *Gravesia*, *Virgatites* and *Zaraiskites*. Early Kimmeridgian faunas of the more northerly province are characterised by *Amoebites*, the more southerly by *Aulacostephanus* and *Rasenia*.

Marine invertebrate biogeography

Faunas of the Tethyan Realm, occupying the rest of the world, have in general a greater diversity than those of the Boreal Realm. Major animal groups such as the lituolacean foraminiferans, tintinnids, radiolarians, hydrozoans and rudist bivalves, together with algae such as dasycladaceans, are largely or entirely confined to the Tethyan Realm, principally to the limestone belt on the southern flank of the Tethys; corals and sponges are much more abundant and diverse.

The ammonites also show greater diversity. The phylloceratids and lytoceratids have long been considered as characteristic of the old Tethyan Ocean, which does not mean that they are abundant in all Tethyan faunas, nor that they do not occur sporadically in the Boreal Realm. The Oppeliidae and Haploceratinae are other characteristic groups from Middle Jurassic times onwards. Besides these, the following ammonite families and sub-families have a predominantly or exclusively Tethyan distribution:

TITHONIAN	Berriasellidae
	Spiticeratinae
	Virgatosphinctinae
KIMMERIDGIAN	Aspidoceratinae
	Ataxioceratinae
	Virgatosphinctinae
	Streblitinae
	Simoceratidae
	Taramelliceratinae
OXFORDIAN	Aspidoceratinae
	Ochetoceratinae
	Taramelliceratinae
CALLOVIAN	Peltoceratinae
	Aspidoceratinae
	Reineckiidae
	Hecticoceratidae
	Macrocephalitidae
BATHONIAN	Sphaeroceratidae
	Clydoniceratidae
	Tulitidae
	Zigzagiceratinae
	Morphoceratidae
BAJOCIAN	Hammatoceratidae
	Leptosphinctinae
TOARCIAN	Hammatoceratidae
	Bouleiceratinae

PLIENSBACHIAN	Hildoceratidae
	Dactylioceratidae
SINEMURIAN and HETTANGIAN	Juraphyllitidae
	Ectocentridae

Distinctive Tethyan belemnites include, from Callovian times onwards, *Belemnopsis, Hibolithes* and *Conodiocoelites*, together with *Duvalia* in the Tithonian, although the first representatives of the Duvaliidae appear in the Toarcian of the Arctic (Stevens 1965, 1973). The related coleoid *Atractites*, which ranges up into the Liassic, is another Tethyan element.

Certain bivalves are restricted to the shallow water limestones on the southern flanks of the western Tethys. They include, in the Liassic, *Lithiotis* and such heavy-hinged genera as *Opisoma, Daharina* and *Gervilleioperna*. The Upper Jurassic rudist *Diceras* is also a characteristic Tethyan genus.

Discohelix (Wendt 1968), *Nerinea* (Imlay 1965, Ziegler 1966) and *Purpuroidea* are three characteristically Tethyan gastropod genera. There are almost certainly many others but the state of gastropod taxonomy leaves much to be desired and does not, as yet, warrant broad generalisations.

Among the brachiopods Ager (1967) has cited the curious family of Pygopidae as the most typical Tethyan forms, but the rare occurrence of *Pygope* in late Jurassic beds of East Greenland should be noted. Other European and North African Tethyan brachiopods include costate terebratuloids (e.g. *Hesperithyris* in the Liassic and *Flabellithyris* and *Eudesia* in the Middle Jurassic), sulcate terebratuloids, axiniform *Zeilleria*, the rhynchonellids *Cirpa* and *Prionorhynchia* and some smooth *Spiriferina*, e.g. *S. alpina*.

The shallow water carbonate belt on the southern side of the Tethys, from the circum-Mediterranean region to the Middle East, contains a characteristic foraminiferan fauna of arenaceous forms with complex internal structure, notably the Lituolidae, Pavonitidae and Dicyclinidae. Typical genera include *Orbitopsella* and *Lituosepta* (Lower Jurassic), *Kilianina, Meyendorffina* and *Orbitammina* (Middle Jurassic), and *Kurnubia* and *Pseudocyclammina* (Upper Jurassic) (Gordon 1970, Hottinger 1971).

Provinces

The Tethyan faunas are dominantly cosmopolitan in distribution, even at the species level, but there are more indications of provinciality than in the case of the Boreal Realm, a fact which may be partly related to the much greater geographic extent of the Tethyan Realm.

Attempts have been made, on the basis of Middle and Upper Jurassic ammonite distributions, to distinguish a Sub-mediterranean from a Mediterranean Province. Kimmeridgian and Tithonian faunas in the former, which extends from southern Poland through southern Germany and France to Portugal, show a general dominance of oppeliids and perisphinctids over phylloceratids, whereas the reverse is true of the Mediterranean Province (Geyer 1961, Ziegler 1963, Zeiss 1968). Thus, in the Lower Tithonian the Sub-mediterranean 'Province' is characterised by a variety of genera such as *Hybonoticeras*, *Gravesia*, *Taramelliceras*, *Lithacoceras*, *Sutneria*, *Aspidoceras* and *Glochiceras*, with very subordinate phylloceratids and lytoceratids, whereas the more southerly Mediterranean 'Province' is more restricted in variety, with only a few genera including *Simoceras*, *Berriasella* and *Haploceras* accompanying abundant phylloceratids and lytoceratids (Zeiss 1968). There is, however, no simple latitudinal relationship and the different faunas may coexist at different horizons in the same area. This throws into some doubt the genuineness of the 'provinciality'. Cariou (1973) has recognised a comparable pattern of ammonite distribution in the Callovian and Oxfordian.

Ager (1967) has also recognised regional variations in Liassic (especially Pliensbachian) brachiopod faunas of the same general region. A relatively restricted Mediterranean 'Province', including the southern and eastern Alps, the peri-Adriatic region, southeastern Spain and North Africa (i.e. the southern flank of the Jurassic Tethys), is characterised by sulcate terebratulids, axiniform *Zeilleria*, *Propygope* and a dominance of *Prionorhynchia* and smooth spiriferoids. Some Tethyan groups are less restricted, however, and extend into northwest Europe. These include cynocephalous rhynchonellids, *Prionorhynchia* and *Cirpa*, which are subordinate elements in the more northerly faunas. According to Stevens (1965, 1973) the European part of the Tethys in the Oxfordian and Kimmeridgian is distinguished from other regions by the presence of *Hibolithes* and the absence of *Belemnopsis*.

Though the bulk of the faunas of East Africa, Madagascar, the Middle East and the northwest part of the Indian Subcontinent closely resemble the better-known European faunas even at species level, there are probably enough fossils largely or entirely confined to these regions to warrant the term 'Ethiopian Province'. Thus the Lower Toarcian ammonite genus *Bouleiceras* is common to Madagascar and Saudi Arabia and also occurs (perhaps abundantly) in Kenya and Pakistan. Some migration into the European Tethys is indicated by its rare occurrence in Morocco, Portugal and Spain (Blaison 1968). Recently its occurrence also in the Lower Toarcian of Chile and Argentina has been reported by Hillebrandt (1973a), so that this genus is not such a good indicator of the province as was formerly thought. The Bajocian ammonite *Ermoceras* is common in Arabia and Sinai but has also been found in Morocco, as has the East African gastropod *Africoconulus* (Cox 1965). More strictly confined to the Ethiopian Province are the Callovian ammonite genera *Obtusicostites* and *Sindeites* (East Africa, Madagascar, Cutch, in northwest India) and the genus *Indogrammatodon*, abundant in the Callovian of East Africa, Arabia and Cutch (Cox 1965). A distinctive Bathonian bivalve fauna with *Eligmus rollandi* and *Gryphaea* (*Africogryphaea*) *costellata* ranged through East Africa and the Middle and Near East and extended to North Africa.

A distinctive group of belemnites, *Belemnopsis* of the *orientalis–gerardi* group, first appeared in the Bathonian and Callovian of Madagascar, and by Oxfordian times had spread into Somalia and India. Further indigenous elements among the belemnites arose in the Kimmeridgian (Stevens 1965, 1973). At the same time the bivalve fauna became more clearly differentiated (Cox 1965). *Indotrigonia* (abundant) and *Opisthotrigonia* (present) occur only in East Africa and India and the same distribution applies to *Astarte sowerbyana* and *Stegoconcha gmuelleri*, while *Indogrammatodon* continued to flourish.

Evidence of indigenous elements is not confined to the molluscs: *Septirhynchia* is a giant pentamerid-like Upper Jurassic rhynchonellid (sufficiently distinct according to Ager (1967) to have its own family) which until recently had only been found in East Africa and Sinai, but which is now known also from Tunisia (Dubar 1967) and southern Iran (James and Wynd 1965).

A number of fossils occurring quite abundantly in the Ethiopian Province are distributed more widely in the Indo-Pacific region but are apparently absent from Europe. Not enough is yet known of the Indo-Pacific faunas and their relationship with the European to justify a clear designation of a province or provinces but the facts of distribution of a number of the more notable molluscs need to be recounted.

Arkell (1956) cited as the best evidence for his Pacific Realm the distribution of the Bajocian ammonites *Pseudotoites* and *Zemistephanus*. Abundant *Pseudotoites* occur in Western Australia and were recorded also from the Moluccas, Argentina, British Columbia and Alaska, but it is apparently absent from the better-known Bajocian deposits of Europe. The Moluccas record is based on one figured specimen and must be considered somewhat doubtful, while the supposed *Pseudotoites* of Canada and Alaska is now assigned to *Zemistephanus* by Imlay (1964). *Zemistephanus* is much rarer in Western Australia but is apparently well authenticated in western Canada and Alaska (Imlay 1964).

Marine invertebrate biogeography

Other indications of ammonites with an Indo-Pacific distribution are limited to the Upper Jurassic. The Mayaitidae are a distinctive Oxfordian family which occurs commonly in the Ethiopian Province and is also found in the Himalayas and Indonesia; perhaps also in Argentina, where *Mayaites* has been recorded but not figured by Stipanicic (1966). A number of Kimmeridgian and Tithonian perisphinctids have a somewhat similar distribution. *Uhligites* occurs in the Ethiopian Province, Turkey, the Himalayas, Indonesia, New Zealand and central and western North America; and *Blanfordiceras* in Pakistan, the Himalayas, Indonesia and South America. *Epicephalites*, *Subneumayria* and certain species of *Idoceras* are found only in Mexico and New Zealand, while *Substeuroceras* occurs along the length of the Western Cordillera of the Americas, the Middle East and Japan. *Paraboliceras* has only been recorded from the Himalayas, Indonesia and New Zealand. Enay (1972, 1973) recognised affinities of the Andean Tithonian faunas with the Mediterranean (e.g. *Durangites*, *Hybonoticeras*, *Pseudolissoceras*) and with the Ethiopian Province (*Blanfordiceras*, *Lytohoplites*). Affinities with countries on the western side of the Pacific are much weaker and appear to preclude a direct migration route.

According to Stevens (1965) an Indo-Pacific Province centred on New Zealand and Indonesia, signified by *Belemnopsis* of the *uhligi* complex and *Conodicoelites*, became clearly differentiated among the belemnites in Kimmeridgian times, extending from the Ethiopian Province and the Himalayas to New Zealand, West Antarctica (Stevens 1967) and South America. The Mediterranean Kimmeridgian belemnite assemblage consisted predominantly of *Hibolithes*. This clear differentiation was lost or at least reduced in the Tithonian.

Among the Bivalvia there appears to be a distinctive group of Upper Jurassic *Buchia* species in the Indo-Pacific region. Thus the Lower Tithonian *B. plicata–hochstetteri* group is found in abundance both in New Zealand and the Himalayas and affords one of the best means of correlation between these regions (Stevens 1968). Jeletsky (1963) has created a new genus, *Malayomaorica*, for a distinctive Kimmeridgian form, formerly grouped with *Buchia*, which is confined to Indonesia, Australia and New Zealand. *Inoceramus galoi* has a geographic range extending from New Zealand and Indonesia, where it is common (Stevens 1968), to Japan (Hayami 1961) and South America. The Liassic genus *Otapiria*, formerly thought to be confined to New Zealand and New Caledonia, is now known also from northern Alaska, northeastern Siberia (Imlay 1967), and Colombia (Geyer 1972), leaving the Liassic *Pseudaucella* as the only possible indigenous mollusc in New Zealand (Fleming 1962). The curious pectinid genus *Weyla* occurs quite commonly in the Ethiopian Province, in Madagascar and East Africa (Cox 1965), and in western North and South America (Hallam 1965), and occurs rarely in Morocco (Dubar 1949), but is not known from anywhere else.

PROVINCIALITY IN RELATION TO PLATE MOVEMENTS

If we accept the geological and geophysical data cited in chapter 7 as indicating that only a small part of the present Atlantic Ocean existed during the Jurassic, then we should expect the neritic invertebrate faunas of the Old and New Worlds to be much more similar to each other than they are today. This expectation is amply borne out by the palaeontological data, and, more generally, the pattern of Jurassic provinciality bears little relation to the present distribution of continents and oceans.

Within the Boreal Realm the Old and New World ammonite faunas are strikingly alike, and at least some of the few purportedly indigenous North American genera are probably invalid taxa or at least relegatable to subgenera. Likewise, only a very small percentage of the American Tethyan Realm genera are indigenous, and further taxonomic work is likely to whittle down rather than increase the number (Hallam 1971*b*). Among the Bivalvia, the same pattern probably applies, though much less taxonomic work has been done on this group. *Plicatostylus* (Liassic of Oregon and Peru) and *Lupherella* (Liassic of Oregon) are the only two recorded indigenous genera, to which one can add *Gryphaea nebraskensis* of the Callovian of the US Western Interior, quite distinct from contemporary European species of *Gryphaea*, and probably one or two other *Gryphaea* species. Further work might well reveal more indigenous species, and it is worth noting that Jeletsky (1965) differed from other workers in considering that North America has its own distinct species of *Buchia*. It is, however, unlikely that differences with the Old World will show up as more than minor.

Within the Old World, the striking similarities of the faunas of the northwestern part of the Indian subcontinent (Cutch and Pakistan) and East Africa–Madagascar, and their relative distinctness from other regions, is inexplicable in terms of present-day geography, and a reconstruction such as that of fig. 10.2 makes much more sense in helping to explain the Ethiopian Province.

The Toarcian to Bajocian ammonite fauna of the Badamu Formation in east-central Iran is intriguing in that it has strong affinities with northern Europe rather than the Mediterranean or Ethiopian provinces of the Tethyan Realm (Seyed Emami 1971). For instance, in the Aalenian

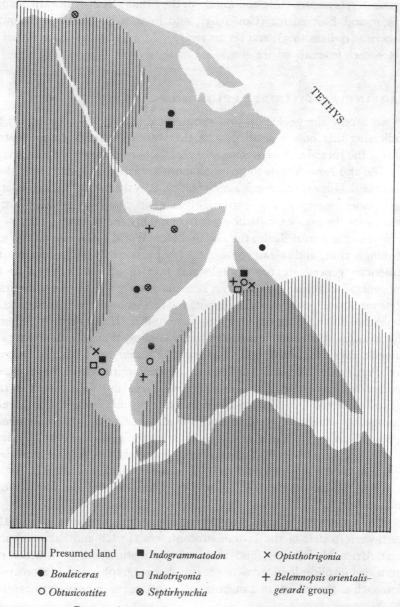

FIGURE 10.2 Proposed pre-drift reconstruction of the East African, Arabian, Indian and Madagascan regions, sharing the site of the Jurassic shallow marine gulf and the distribution of some distinctive 'Ethiopian' elements of the fauna (Hallam 1971*b*).

graphoceratids are dominant, rather than hammatoceratids as in the Mediterranean faunas. As noted in chapter 6, this evidence is consistent with a Jurassic position for Iran north of the Zagros Crush Zone on the northern side of the Tethys. The evidence is not strong enough to be more than suggestive, and more work is badly needed on other Jurassic Iranian faunas. Independent support for a relatively northerly position is provided, however, by the distribution of coals and evaporites in Iran, as discussed in chapter 9.

Finally, the apparently strong affinities of the Indonesian molluscan faunas with Australasia, noted earlier in the chapter, could be held tentatively to support a position for the islands of southeast Asia on the southern side of the Tethys (see chapter 7), but much more work on the distribution of the Asian faunas is needed, for instance to determine how different the faunas of Japan are from those of Indonesia and Australasia. Strong affinities with the Western Tethys region are recognisable in a Toarcian bivalve fauna from South Vietnam, recently described by Hayami (1972).

Rearranging Jurassic continents on the globe manifestly fails to explain, however, either the differentiation of the Boreal and Tethyan realms, or the Ethiopian Province within the latter, because free shelf-sea communication around the western margins of the Tethys must have existed throughout most if not all of the period. Even at the close of the Jurassic any deep oceanic barrier between Laurasia and Gondwanaland must have been narrow and hence unlikely to prevent free cross-migration of planktonic larvae and nektonic adults.

CAUSES OF PROVINCIALITY

Interpreting the Boreal–Tethyan provinciality has been the subject of much controversy, which is likely to continue for some time yet in view of the complexity of the subject and insufficiency of 'hard data' of the right sort. The various explanations which have been put forward can be discussed under five subheadings.

Temperature

Some form of temperature control was proposed by Neumayr (1883) and Uhlig (1911) and remains the most popular interpretation among current workers (e.g. Sato 1960, Ziegler 1964, Saks *et al.* 1964, Jeletsky 1965, Donovan 1967, Stevens 1971, Enay 1972, 1973, Scheiberovna 1972). The evidence proposed in favour is, firstly, that there is a northward decrease in diversity from the Tethyan to the Boreal Realm, especially among reef corals, suggesting a comparison with climatic zonation at the present day in the Northern Hemisphere, and secondly, there is a concomitant marked decline in the abundance of limestones northwards from southern Europe. Additional support from oxygen isotope determinations has been put forward by Stevens (1971) but, bearing in mind the critique of the results given in the last chapter, this line of evidence is best disregarded.

A number of objections may be raised to a simple hypothesis of temperature control. In the first place, we have seen in chapter 9 how the evidence of terrestrial organisms overwhelmingly favours an equable climate, with a much more modest decline of temperature with increasing latitude than at the present day. The marine environment is well known to be more equable than the continental, yet quite drastic changes in Jurassic marine faunas of Europe take place within a few degrees of latitude. The occurrence of reef corals in high Jurassic latitudes in eastern Asia suggests that their absence from similar or lower latitudes elsewhere is attributable to other factors. More generally, temperature reduction or some other latitude-dependent factor is by no means the only way to account for diversity reduction.

It was also argued in chapter 9 that the presence or absence of limestones is primarily a function of the tectonic-sedimentational regime and in equable periods like the Jurassic substantial carbonate deposits could have accumulated in most latitudinal zones if temperature were the principal controlling agent. The sparsity of calcareous material in the highly condensed Volgian deposits of the Russian Platform, in an environment where limestones might be expected to have accumulated by default, as it were, of terrigenous clastic sedimentation, is readily explicable if it is assumed that late Jurassic limestones are primarily organic in origin. The contemporary deposits of the Tethys are rich in coccoliths, which, being pelagic, might have been excluded from the shallow seas of continental interiors by a variety of ecological factors.

If one is to apply an actualistic model to the Jurassic, then some degree of latitudinal bipolarity might have been expected, yet no Austral faunas have been convincingly demonstrated to match the Boreal. Gordon (1970) claimed a degree of bipolarity for the foraminiferan faunas, because the Middle and Upper Jurassic faunas of Sinai and Somalia tend to resemble those of northern Europe, rather than those of the Tethyan belt running from the western Mediterranean to Arabia. However, on the Jurassic reconstruction portrayed in this book Sinai actually lies in a latitudinal zone to the north of Arabia!

A final point is that modern work has undermined any straightforward temperature control hypothesis for latitudinal diversity reduction even at the present day (Valentine 1972). It turns out that many deep sea communities are highly diverse compared with comparable shelf-sea communities, although the former live at considerably lower temperatures.

Physical barriers

A smaller number of workers (e.g. Uhlig 1911, Arkell 1956, Imlay 1965) have invoked some sort of physical barrier, such as a landmass, as at least a contributary factor. Such an interpretation may be objected to on several grounds:

(i) The changes from one realm to another are gradational in character and there are many faunal elements common to both. This is not compatible with what is known about physical barriers in the modern oceans.

(ii) The boundary between the two realms oscillated geographically with time, as between the Callovian–Lower Oxfordian and Upper Oxfordian.

(iii) Arkell's invocation of a land barrier, to isolate the Arctic Ocean and hence allowing the development of a distinctive ammonite fauna in the Callovian, has been rendered less plausible by the discovery of a rich boreal fauna of Bathonian Cadoceratinae in the Arctic.

(iv) The hypothesis suffers from vagueness. The physical barriers have not usually been clearly specified nor independent evidence sought for their existence. A modified version of the hypothesis invokes ocean currents as well (Imlay 1965). This also is of doubtful significance, because the zoogeographic influence of ocean currents today is primarily one of temperature, which seems not to have been a dominant control in the Jurassic.

Stevens (1965, 1973) suggested barriers of deep sea within the Tethyan Realm to account for the late Jurassic differentiation of belemnite faunas, on the assumption that the organisms were stenothermal and confined to shelf sea habitats, like many living squids and cuttlefish.

Depth of sea

Depth is not a primary environmental factor but it often correlates with other more significant factors such as food supply, temperature, pressure and incidence of light. There is a widespread notion, dating back to Haug (1907), that the phylloceratids occupied deeper water than other ammonites, hence accounting for their restriction in Europe to the Mediterranean–Caucasus belt, where rocks of deeper water facies occur. There seem to be quite good reasons for accepting this as a plausible explanation of the contrast between the Mediterranean and Sub-mediterranean ammonite provinces, which are therefore not good 'provinces'. Likewise, phylloceratids did not extend into the shallow seas of the US Western Interior, but are quite abundant in the deeper water, more open sea Alaskan deposits far to the north.

Ziegler (1967) has gone so far as to suggest a sevenfold depth zonation for the late Jurassic of southern Europe based on ammonites and other groups. Among the ammonites, aspidoceratids and perisphinctids with subordinate oppeliids occupies the shallowest zone. An intermediate zone is marked by a dominance of oppeliids over the other two groups, while the deepest zone where megafossils occur in reasonable abundance was dominated by phylloceratids and lytoceratids. Ziegler put actual values on the depth zones in terms of metres, but this must be regarded as very speculative, although the figures cited are not unreasonable. It ought to be recognised, of course, that even if the phylloceratids were more restricted than other groups to a pelagic environment they need not necessarily have *lived* in deeper water, being nektonic organisms, but simply might have dropped to the bottom after death in a more offshore and hence, in most cases, deeper sea.

There can be no serious suggestion that significant depth differences controlled the differentiation into Tethyan and Boreal realms.

Salinity

Viewing the world as a whole, no simple relationship is perceptible between faunal provinciality and sedimentary facies, except that certain groups of organisms such as the lituolacean foraminiferans and certain bivalves including rudists are more or less restricted to the shallow water carbonate facies on the southern flanks of the western Tethys. Within Europe, however, the northward transition from carbonate to terrigenous clastic and ferruginous facies corresponds closely with the boundary of the Tethyan and Boreal Realms. Furthermore, spatial fluctuations of this boundary with time find an echo in the sedimentary change. This correlation may break down in detail, but the broad relationship is undeniable and clearly demands an explanation.

A reasonable interpretation seemed to be that the Boreal Realm in Europe occupied a shallow epeiric sea whose salinity was lowered sufficiently, by an influx of fresh water from rivers draining a humid hinterland, to exclude some of the most strictly stenohaline organisms. A condition of substantial palaeogeographic stability ensured that some slightly more euryhaline organisms were able to evolve independently in the relative absence of competition and give rise to a provincial fauna. This explanation had the advantage of accounting for diversity reduction northwards from the Tethyan Ocean, as well as the increased faunal abundance or density in the Boreal Realm, especially among the Bivalvia.

Once a boreal fauna was established, biological competition with Tethyan invaders could have helped ensure its survival (Hallam 1969*a*, 1971*b*). It remained difficult to see, however, how a sea with a mean salinity only subtly different from that of the ocean (because otherwise cephalopods, brachiopods, and echinoderms would have been excluded, as manifestly they were not) could have survived in more or less the same region for tens of millions of years. That actualistic analogues are hard to find is another, but less serious, objection, because epeiric seas extending far into continental interiors are a phenomenon of the past rather than the present.

The weakness of all the hypotheses reviewed so far is that they concentrate too extensively on a single environmental factor and do not take adequately into account the very considerable advances made in recent years in our ecological understanding. Furthermore there has been an absence of intensive work devoted to collecting detailed data of the sort required to test satisfactorily one hypothesis against another. An attempt to do this for the Pliensbachian and Toarcian stages over a 40° range of latitude from Greenland to Morocco has led to a new interpretation (Hallam 1972*a*) which will be outlined below.

Environmental stability

The Pliensbachian and Toarcian are highly suitable for analysis since their stratigraphy and much of their palaeontology are comparatively well known and the former stage at least illustrates well the division into Tethyan and Boreal realms among the ammonites. In the ecological analysis which was undertaken attention was confined to the molluscs and articulate brachiopods which constitute by far the larger part of the invertebrate macrofauna. A series of regions were distinguished over a wide range of latitude, with the sedimentary facies passing from terrigenous clastic and ferruginous deposits in the north to limestones and marls in the south (fig. 10.3).

The key faunal data sought are *diversity*, expressing the variety, and *density*, expressing the individual abundance. *Within-region* diversity changes relate to variations in local environmental factors such as depth, as discussed in chapters 3 and 4. *Between-region* diversity changes should give the regional 'signals' as opposed to the local 'noise', provided enough deposits are sampled within a given region and the maximum values of diversity compared.

Diversity can be analysed at any taxonomic level, and whereas most ecological studies have concerned themselves with species, in this case it is more practical to deal with genera. While Jurassic taxonomists have by and large achieved a high measure of accord about generic distinctions, this is decidedly not the case with species, where considerable confusion often reigns. It is pertinent to enquire how much information may be lost by restricting attention to genera.

Jurassic ammonites have been oversplit by typologically-minded taxonomists, to the extent that widely recognised genera may often correspond more or less to broadly interpreted Linnaean species and the so-called species are no more than morphological variants. The bivalve, gastropod and brachiopod species commonly range through one or more stages. Within the confines of a substage in a given region, genera are mostly represented by only one species, and most of the remaining genera by only one common species. Therefore generic and specific analysis by substage should give closely similar results, as they do for a world-wide diversity analysis of Recent bivalves (Stehli *et al.* 1967). It is believed, therefore, that little essential information is lost by ignoring specific distinctions.

The study area was divided into four regions, from north to south: Greenland, England, the Iberian peninsula and Morocco (fig. 10.3). England was subdivided into northeast and southwest areas to provide some control on variations over a relatively short distance which nevertheless exhibits a moderate facies change in the Pliensbachian, from more calcareous sediments in the southwest to more sandy and ferruginous ones in the northeast. The stratigraphical intervals chosen were Lower and Upper Pliensbachian and Middle Toarcian.

The results of the generic diversity analysis are given in Hallam (1972*a*, figs. 3 to 6). There is no simple pattern of north–south diversity change from north to south common to all the groups analysed (ammonites, bivalves, gastropods and brachiopods). Whereas the Lower Pliensbachian and Toarcian ammonites show a diversity reduction northwards, the Upper Pliensbachian picture is complicated by the existence of the boreal amaltheid fauna. The brachiopods and gastropods generally reveal no notable change in either direction, though the brachiopods tend to show a modest northward diversity reduction (the Upper Pliensbachian gastropods are a somewhat special case related to the rich fauna of the Moroccan reefoid deposits). In three cases out of four the bivalves show a northward *increase* in diversity and in the fourth case only a negligible northwards diminution. The bivalve data alone are sufficient to throw into question both the temperature and salinity control hypotheses.

According to Stehli *et al.* (1967) the diversity of Recent bivalves shows a significant reduction as latitude increases away from the tropics, both at the generic and specific level. Across the 40° range of latitude separating the

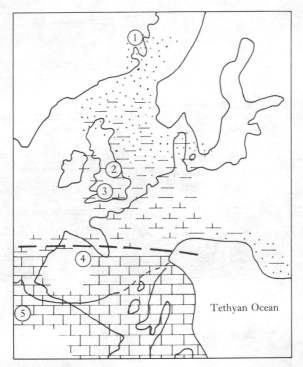

FIGURE 10.3   Regional changes in sedimentary facies for the Pliensbachian of the North Atlantic–West Tethyan region. Bold broken line signifies northern limit of Tethyan faunal realm as marked by abundant hildoceratid and dactylioceratid ammonites. Numbers refer to sampling stations (Hallam 1972*a*).

northernmost and southernmost areas of the study, generic diversity in the present Atlantic is reduced approximately threefold, and even more over a similar range in the Pacific. These changes are so pronounced that, even in a more equable climate than today, a northward reduction in diversity seems essential to support any temperature control hypothesis.

With regard to salinity, good data are available from the Baltic Sea. Species diversity is reduced almost threefold from the Kattegat (salinity 35‰) to the Belt Sea (salinity 25–30‰) (Segerstråle 1957).

It is far more difficult to obtain valid quantitative data on faunal density. In particular, allowance must be made for variation in rates of sedimentation. The highly schematic portrayal of major density changes given in fig. 10.4 is thought, however, to give a reasonable indication of the principal changes from north to south.

Our interpretative model must account for a general but sometimes slight northward reduction in diversity of the ammonites and brachiopods,

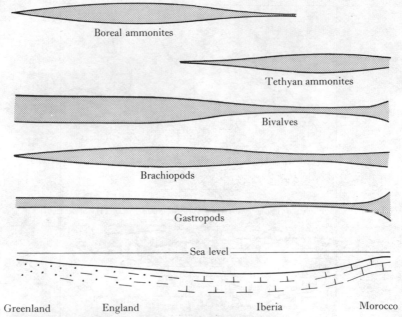

FIGURE 10.4   Highly schematic portrayal of regional changes in Pliensbachian faunal density (Hallam 1972*a*).

together with other fossil groups not considered here, with some tendency to an opposite trend in the bivalves, coupled with a general northward increase in faunal density, especially among the bivalves. It must also account for the associated changes in the sediments which suggest a northward passage from southern Europe from an offshore to an inshore facies (Hallam 1969*a*, 1971*b*; see chapters 3 to 5).

The key concept required is that of environmental stability (Sanders 1968; Bretsky and Lonrenz 1971; Woodwell and Smith 1969). The discovery that benthonic species diversities in the deep sea were higher than in inshore zones of the continental shelf led to the formulation by Sanders (1968) of the stability–time hypothesis. Sanders showed that, for the groups he studied (polychaetes and bivalves), species–individual diversities in the modern ocean diminished in the sequence: tropical shallow water, continental slope, outer shelf, tropical estuary, boreal inner shelf, and boreal estuary. This is an order of decreasing physical stability or predictability of the environment in terms of fluctuating temperature or salinity. Faunas in the unstable environments are subject to high stress and accordingly restricted to eurytopic organisms of comparatively low diversity. Faunas in the stable environments have, over the course of time, evolved a high diversity of stenotopic, biologically accomodated organisms with narrow niche space. Faunal density is normally higher in regions of low stability. The species–individual diversities estimated by Sanders correlate well with species diversities comparable to those used by Hallam (1972*a*).

The Liassic patterns may be explained adequately in terms of modern ecological concepts by postulating a northward reduction in environmental stability from the Tethyan margins to the inshore waters of northern Europe and Greenland, into which a number of rivers debouched from a northern landmass. Offshore to inshore diversity reduction, allied with density increase in the same direction, is well documented in a number of examples from the North American Palaeozoic, as reviewed by Bretsky and Lorenz (1971) and Stevens (1971). Furthermore, Bretsky and Lorenz cite evidence of bivalves being relatively eurytopic and therefore less subject to the diversity reduction inshore. In fact a common Palaeozoic pattern is for bivalves and gastropods to be more diverse inshore and brachiopods more diverse offshore. In the Cretaceous of the American Western Interior, furthermore, bivalves occur in higher density and diversity in the terrigenous clastic facies in the west than in the calcareous facies to the east (Kauffman 1967).

The reduction in stability might have operated in more than one way. The sedimentary facies patterns suggest influx of fresh water from northern lands, which would effect some degree of salinity reduction close to the coast. In so far as the change is more or less from south to north, stronger seasonal contrasts in temperature in the same direction could also act to reduce stability. Perhaps the most important factor, however, was the general reduction in water depth, because an extensive sheet of very shallow water would be subject to greater and relatively more unpredictable variations in such factors as salinity, temperature and perhaps content of dissolved oxygen than the deeper sea further offshore.

Since the regional changes outlined above are in many respects typical of the Tethyan–Boreal provinciality in Europe for the Jurassic as a whole, a number of wider implications may be discussed.

Evidently the ammonite and brachiopod faunas were on the whole more stenotopic than the bivalves. Within the ammonites, the boreal genera emerge from the analysis as eurytopic derivatives of Tethyan stocks which evolved independently as a consequence of protective isolation by the ecological barrier of physical instability. The fact that 'Tethyan' hildoceratids and dactylioceratids did not invade the Boreal Realm in large quantities until after the extinction of the amaltheids suggests that perhaps biological competition also played a role. A truly

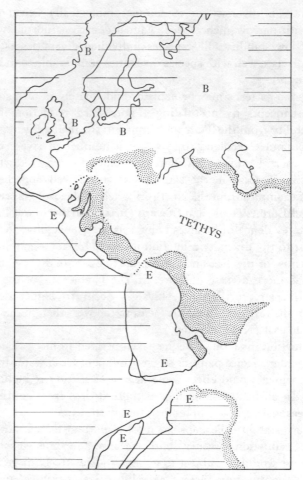

FIGURE 10.5 Jurassic faunal provinces in relation to the Tethys ocean. Stippled areas signify Alpine fold belt lands of uncertain position; horizontal lines signify presumed land at time of maximum extent of Jurassic seas. B, Boreal Realm; E, Ethiopian 'Province' (Hallam 1972a).

boreal ammonite fauna was not re-established until the late Bajocian or Bathonian.

One of the most interesting implications of the study is that faunal provincialism can develop in epicontinental seas marginal to the ocean without the necessary intervention either of topographic barriers, in the form of land or wide stretches of deep water, or of temperature gradients. The Ethiopian Province of the Tethyan Realm is an excellent example. On the reconstruction adopted in this book this is seen to occur in an epicontinental sea on the southern side of the Tethys (fig. 10.5). Free shelf-sea communication between Europe and Africa would have been possible around the western margin of the Tethys, at least until very late in the period, and even then the oceanic separation would have been very narrow, and no serious obstacle to cross migration. There could have been no appreciable temperature difference, furthermore, between southwest Europe and Arabia.

A striking parallel exists between this interpretation and that of Palmer (1973) to account for the distribution of Cambrian trilobites. He recognises several provincial faunas which were associated with the deposits of epicontinental seas. The Agnostida were widely distributed in high diversity faunas in areas with free connections to the open ocean. Periodic extinction of the endemic faunas was followed by renewal from the 'pelagic' stock.

A good deal of confusion exists about the role of water temperature in controlling faunal distributions. Though widely invoked as the dominant regional control on Tethyan and Boreal marine faunas, there has been little discussion as to precisely what was implied. It would have been reasonable, for instance, to examine for Recent faunas the differing effects of minimum, maximum and mean annual temperatures and the different tolerances of larvae as opposed to adults. In fact the modern discovery of high diversity faunas existing at low temperatures on the deep sea floor effectively rules out, as noted already, a simple temperature control as a general explanation of diminishing diversity with increasing latitude. Other latitude-dependent factors, such as the amount of daily illumination and its bearing on phytoplankton productivity, should be further explored, and in fact Valentine (1971, 1972) has argued that trophic resource stability may be the fundamental controlling factor.

It is worth noting that Brookfield (1974), in his account of late Oxfordian and early Kimmeridgian facies in southern England, has also adopted an interpretation of faunal provinciality in terms of 'stress' factors related to environmental stability.

Invoking environmental stability as perhaps the major factor controlling the regional differentiation of Jurassic faunas has the advantage both of explaining more facts than other hypotheses and taking into account modern ecological concepts and knowledge. This is not to say that the hypothesis has no weaknesses. Environmental stability is in itself a somewhat vague concept in certain respects and needs examining more rigorously from a variety of points of view. We should like to know how much more eurytopic and tolerant of inshore environments the Recent bivalves may be in comparison to other groups in seas at the present day, for instance. Obviously many more detailed studies of Jurassic faunas in relation to plausible environmental models are required. Studies of Jurassic faunal provinciality are still in a preliminary phase, and we may confidently anticipate within the next decade or so advances in our understanding as significant as those of the last decade in the field of sedimentary environments.

Epilogue

The Jurassic world that can be reconstructed was very different from the one we have direct experience of at the present day. In the absence of polar ice caps the climate was appreciably more equable, with a warm climate extending to quite high latitudes. Mainly as a consequence of this, and the greater coherence of the continental areas, with only one major ocean in existence, both faunas and floras exhibited far less provinciality and diversity, in addition, of course, to being more primitive. Though mountains undoubtedly existed, notably around parts of the Pacific margins, continental relief was more modest, with shallow epeiric seas occupying a much greater extent. These seas were quite possibly almost tideless over large regions, and free of the effects of deep oceanic circulation. As a consequence of the greater equability it is also likely that the circulation of oceanic waters and winds was more sluggish. Tectonism, moreover, does not appear to have been significant on more than a local scale until the end of the period. This late Jurassic increase in tectonic activity can be regarded as a prelude to the much more extensive disturbances in the Cretaceous, by the end of which period many of the principal features of the present world configuration had been established. Thus the Jurassic could be held to mark the end of an *ancien régime*.

Nearly three centuries ago Thomas Burnet attempted to portray, in his *Sacred Theory of the Earth*, his vision of the Earth before the Deluge. Though long since relegated to the realm of the fanciful, his words might be thought to give a not too outrageously distorted overall impression of the Jurassic world:

It had the Beauty of Youth and blooming Nature, fresh and fruitful, and not a Wrinkle, Scar or Fracture in all its Body; no Rocks nor Mountains, no hollow Caves, nor gaping Channels, but even and uniform all over. And the Smoothness of the Earth made the Face of the Heavens so too; the Air was calm and serene; none of those tumultuary Motions and conflicts of Vapours, which the Mountains and the Winds cause in ours; 'Twas suited to a golden Age.

References

Agard, J. and du Dresnay, R. 1965. La region minéralisée du J. Bou Dahar, près de Beni-Tajjite. *Notes Mém. Serv. géol. Maroc.* No. 181, 135–152.

Ager, D. V. 1967. Some Mesozoic brachiopods in the Tethys region. *Systematics Assoc. Publ.* No. 7, 135–150.

Ager, D. W. and Evamy, B. D. 1963. The geology of the southern French Jura *Proc. Geol. Ass. Lond.* **74**, 325–356.

Ager, D. V. and Wallace, P. 1966. The environmental history of the Boulonnais, France. *Proc. Geol. Ass. Lond.* **77**, 385–417.

Ager, D. V. and Wallace, P. 1970. The distribution and significance of trace fossils in the uppermost Jurassic rocks of the Boulonais, northern France. In: T. P. Crimes and J. C. Harper (editors), *Trace fossils*, Seel House Press, Liverpool, pp. 1–18.

Agrawal, S. K. 1956. Contribution a l'étude stratigraphique et paléontologique du jurassique du Kutch (Inde). *Ann. Centre. Etudes Doc. Paléont.* No. 19.

Aitken, W. G. 1961. Geology and palaeontology of the Jurassic and Cretaceous of southern Tanganyika. *Bull. geol. Surv. Tanganyika* 31, 1–144.

Aldinger, H. 1957. Zur Entstehung der Eisenoolithe im Schwäbischen Jura. *Zeitschr. deutsch. Geol. Ges.* **109**, 7–9.

Alencaster de Cserna, G. 1963. Pelecipodos del Jurasico medio del noroeste de Oaxaca y noreste de Guenero. *Paleont. Mexicana* No. 15.

Allen, P. 1969. Lower Cretaceous sourcelands and the North Atlantic. *Nature* **222**, 657–658.

Allen, P. and Keith, M. L. 1965. Carbon isotope ratios and palaeosalinities of Purbeck–Wealden carbonates. *Nature* **208**, 1278–1280.

Amaral, G., Cordari, U. S., Kawashita, K. and Reynolds, J. H. 1966. Potassium–argon dates of basaltic rocks from southern Brazil. *Geochim. Cosmochim. Acta* **30**, 159–189.

Ambroggi, R. 1963. Etude géologique du versant meridional du Haut Atlas occidental et de la plaine du Souss. *Notes Mém. Serv. géol. Maroc.* **157**.

Andal, D. R., Esguarra, J. G., Hashimoto, W., Reyer, B. P. and Sato, T. 1968. The Jurassic Mansalay Formation, southern Mindoro, Philippines. *Geol. Palaeont. S.E. Asia* **4**, 179–197.

Andelković, M. Z. 1966. Die Ammoniten aus den Schichten mit *Aspidoceras acanthicum* des Gebirges Stara Planina in Ostserbien (Jugoslavien). *Palaeont. Jugoslavica* **6**. (In Serbian, with German summary.)

Anderson, F. W. and Hughes, N. F. 1964. The 'Wealden' of north-west Germany and its English equivalents. *Nature* **201**, 907–908.

Anderson, R. Y. and Kirkland, D. W. 1960. Origin, varves and cycles of Jurassic Todilto Formation. *Bull. Am. Ass. Petrol. Geol.* **44**, 37–52.

Androusev, D. 1965. *Geology of the Czechoslovakian Carpathians, Vol. 2, Mesozoic.* Sloven. Akad. Vied, Bratislava. (In Czech, with Russian and German summaries.)

Arkell, W. J. 1933. *The Jurassic System in Great Britain.* Oxford Univ. Press.

Arkell, W. J. 1935. On the nature, origin and climatic significance of coral reefs in the vicinity of Oxford. *Quart. J. geol. Soc. Lond.* **91**, 77–110.

Arkell, W. J. 1956. *Jurassic geology of the world.* Oliver & Boyd, Edinburgh.

Armstrong, C. F. and Oriel, S. S. 1965. Tectonic development of Idaho–Wyoming thrust belt. *Bull. Am. Ass. Petrol. Geol.* **49**, 1842–1866.

Armstrong, R. L. and Hansen, E. 1966. Cordilleran infracture in the eastern Great Basin. *Am. J. Sci.* **264**, 112–127.

Armstrong, R. L. and Suppe, J. 1973. Potassium–argon geochronometry of Mesozoic igneous rocks in Nevada, Utah and southern California. *Bull. geol. Soc. Am.* **84**, 1375–1392.

Assereto, R. 1966. The Jurassic Shemshak Formation in central Elburz (Iran) *Riv. Ital. Paleont. Stratig.* **72**, 113–182.

Assereto, R., Barnard, P. D. W. and Fantini Sestini, N. 1968. Jurassic stratigraphy of the Central Elburz (Iran). *Riv. Ital. Paleont.* **74**, 3–21.

Aubouin, J. 1959. Contribution a l'étude géologique de la Grèce septentrionale: les confins de l'Epire et de la Thessalie. *Ann. Geol. Pays Helleniques.* Ser. 1, **10**, 1–525.

Aubouin, J. 1964. Reflexions sur le facies 'Ammonitico Rosso'. *Bull. Soc. géol. France* (7), **6**, 475–501.

Aubouin, J., Brunn, J. H., Celet, P., Dercourt, J., Godfriaux, I. and Mercier, J. 1962. *Esquisse de la géologie de la Grèce. Livre mém. P. Fallot Vol. 2,* 583–610. (editor M. Durand Delga), *Soc. géol. France.*

Aubouin, J., Cadet, J. P., Rampnoux, J. P. 1965. A propos de l'âge de la série ophiolitique dans les Dinarides yougoslaves; la coupe de Milhajlovici aux confins de la Serbie et du Montenegro. *Bull. Soc. géol. France* (7), **6**, 107–112.

Audley-Charles, M. C., Carter, D. J. and Milsom, J. S. 1972. Tectonic development of eastern Indonesia in relation to Gondwanaland dispersal. *Nature Phys. Sci.* **239**, 35–39.

Australian Petroleum Company Proprietary Ltd. 1961. Geological results of petroleum exploration in western Papua 1937–1961. *J. geol. Soc. Austral.* **8**, 1–133.

Bailey, E. B. and Weir, J. 1932. Submarine faulting in Kimmeridgian times: East Sutherland. *Trans. roy. Soc. Edinb.* **47**, 431–467.

Bailey, E. H., Irwin, W. P. and Jones, D. L. 1964. Franciscan and related rocks, and their significance in the geology of western California. *Calif. Div. Mines Geol. Bull.* **183**.

Bakker, R. T. 1972. Anatomical and ecological evidence of endothermy in dinosaurs. *Nature* **238**, 81–85.

Ballance, P. F. and Watters, W. A. 1971. The Mawson diamictite and the Carapace sandstone formations of the Ferrar Group at Allen Hills and Carapace Nunatak, Victoria Land, Antarctica. *N.Z.J. Geol. Geophys.* **14**, 512–527.

Barbera Lamagna, C. 1970. Stratigrafia e paleontologia della formazione degli scisti ad aptici dei dintorni di Bolognolo (Macerata). *Mem. Soc. Nat. Napoli, suppl. al Boll.* **78**, 215–244.

References

Barbulescu, A. 1971. Les facies du Jurassique dans la partie ouest de la Dobrogea centrale (Roumanie). *Ann. Inst. Geol. Publ. Hungar.* **54**, 225–232.

Barnard, P. D. W. 1973 Mesozoic floras. In: N. F. Hughes (editor), *Organisms and Continents through Time*. Paleont. Spec. Papers, No. 12, 175–188.

Barthel, K. W. 1969. Die obertithonische, regressive Flachwasser-Phase der Neuburger Folge in Bayern. *Bayer. Akad. Wiss. Math. – Naturwiss. Kl. Abh.* N.S. **142**, 1–174.

Barthel, K. W. 1970. On the deposition of the Solnhofen lithographic limestone (Lower Tithonian, Bavaria, Germany). *N. Jb. Geol. Paläont. Abh.* **135**, 1–18.

Barthel, K. W. 1972. The genesis of the Solnhofen lithographic limestone (Lower Tithonian): further data and comments. *N. Jb. Geol. Paläont. Mh.* **3**, 133–145.

Barthel, K. W., Cediel, F., Geyer, O. F. and Remane, J. 1966. Der subbetische Jura von Cehegin (Provinz Murcia, Spanien). *Mitt. Bayer. Staatssamml. Paläont. hist. Geol.* **6**, 167–211.

Bate, R. H. 1967. Stratigraphy and palaeogeography of the Yorkshire Oolites and their relationships with the Lincolnshire Limestone. *Bull. Brit. Mus. (Nat. Hist.)*, *Geol.* **14**, 111–141.

Bauer, V. 1929. Ueber das Tierleben auf den Seegraswiesen des Mittelmeeres. *Zool. Jb. Abt.* **56**, 1–42.

Beauvais, L. 1973. Upper Jurassic hermatypic corals. In: A. Hallam (editor), *Atlas of Palaeobiogeography*, Elsevier, Amsterdam, pp. 317–328.

Behmel, H. 1970. Stratigraphie und Fazies im präbetischen Jura van Albecete und Nord-Murcia. *N. Jb. Geol. Paläont. Abh.* **137**, 1–102.

Behmel, H. and Geyer, O. F. 1966. Stratigraphie und Fossilfuhrung im Unterjura von Albarracin (Provinz Teruel). *N. Jb. Geol. Paläont. Abh.* **124**, 1–52.

Benda, L. 1964. Die Jura-Flora aus der Saighan-Serie Nord-Afghanistans. *Beih. Geol. Jb.* **70**, 99–152.

Berg, H. C., Jones, D. L. and Richter, D. H. 1972. Gravina–Nutzotin belt – tectonic significance of an Upper Mesozoic sedimentary and volcanic sequence in southern and southeastern Alaska. *U.S. Geol. Surv. Prof.* Paper 800-D, 1–24.

Bernoulli, D. 1964. Zur Geologie des Monte Generoso (Lombardische Alpen). *Beitr. geol. Karte Schweiz*, N.S. **118**.

Bernoulli, D. 1971. Redeposited pelagic sediments in the Jurassic of the central Mediterranean area. *Ann. Inst. Geol. Publ. Hungar.* **54**, 71–90.

Bernoulli, D. 1972. North Atlantic and Mediterranean Mesozoic facies: a comparison. *Init. Rep. Deep Sea Drilling Project* **11**, 801–871.

Bernoulli, D. and Jenkyns, H. C. 1970. A Jurassic basin: the Glasenbach Gorge, Salzburg, Austria. *Verh. Geol, B-A. Wien* **4**, 504–531.

Bernoulli, D. and Peters, T. 1970. Traces of rhyolitic–trachytic volcanism in the Upper Jurassic of the southern Alps. *Ecl. geol. Helv.* **63**, 609–621.

Bernoulli, D. and Renz, O. 1970. Jurassic carbonate facies and new ammonite faunas from western Greece. *Ecl. geol. Helv.* **63**, 573–607.

Bernoulli, D. and Wagner, C. W. 1971. Subaerial diagenesis and fossil caliche deposits in the Calcare Massicio formation. (Lower Jurassic, Central Apennines, Italy.) *N. Jb. Geol. Paläont. Abh.* **138**, 135–149.

Bernoulli, D. and Jenkyns, H. C. 1974. Alpine, Mediterranean and North Atlantic Mesozoic facies in relation to the early evolution of the Tethys. In: R. H.

Dott and R. Shaver (editors), *Geosynclinal sedimentation, Modern and Ancient, a Symposium*, S.E.P.M. spec. Publ. pp. 129–160.

Berridge, N. G. and Ivimey-Cook, H. C. 1967. The geology of a Geological Survey borehole at Lossiemouth, Morayshire. *Bull. geol. Surv. G.B.* No. 27, 155–169.

Bertraneau, J. 1955. Contributions a l'étude géologique des monts du Hodna: 1, le massif du Bon Taleb. *Bull. Serv. Carte géol. Alger. Sér. B*, No. 9.

Besairie, H. and Collignon, M. 1956. Madagascar. *Lex. Strat. Internat.* Vol. 4 (editors D. T. Donovan and J. E. Hemingway), fasc. 11.

Beuther, A., Dahm, H., Kneuper-Haack, F., Mensink, H. and Tischer, G. 1966. Der Jura und Wealden in Nordost-Spanien. *Beih. Geol. Jb.* **44**.

Beznozov, N. V., Mikhailov, N. P. and Tuchkov, I. I. 1962. *Jurassic Geology of the World*. Revised account of Soviet section of Arkell's book. Nat. Lending Library, Boston Spa, England, R.T.S. 4562 (English translation).

Bielecka, W. 1960. Micropalaeontological stratigraphy of the lower Malm of the Chrzanow region, southern Poland. *Proc. Inst. Geol. Poland*, **31**. (In Polish, with English summary.)

Biese, W. A. 1957 a. Zur Verbreitung des marinen Jura im chilenischen Raum der andinen Geosyncline. *Geol. Rundsch.* **45**, 877–919.

Biese, W. A. 1957 b. Der Jura von Cerritos Bayos – Calama Republica de Chile, Provinz Antofagasta. *Geol. Jb.* **72**, 439–485.

Bigarella, J. J. and Salamuni, R. 1964. Palaeowind patterns in the Botucatü Sandstone (Triassic–Jurassic) of Brazil and Uruguay. In: A. E. M. Nairn (editor), *Problems in Palaeoclimatology*, Interscience, London and New York, pp. 406–409.

Birkenmajer, K. and Znosko, J. 1955. Contribution to the stratigraphy of the Dogger and Malm in the Piening klippen Belt (Central Carpathians). *Roczn. Polske Towerz. Geol.* **23**, 3–36. (Polish, with English summary.)

Bitterli, P. 1963 a. Aspects of the genesis of bituminous rock sequences. *Geol. Mijnb.* **42**, 183–201.

Bitterli, P. 1963 b. Classification of bituminous rocks of Western Europe. *Proc. 6th World Petrol. Congr.* Sect. 1, Paper 30.

Black, M. 1929. Drifted plant-beds of the Upper Estuarine Series of Yorkshire. *Quart. J. geol. Soc. Lond.* **85**, 389–437.

Blaison, J. 1961. Stratigraphie et zonéographie du lias inférieur des environs de Lons-le-Saumier, Jura. *Ann. Sci. Univ. Besançon*, Ser. 2, *Geol.* **15**, 35–122.

Blaison, J. 1963. Observations nouvelles sur la stratigraphie du Jurassique de la region de Kandreho (Madagascar). *Bull. Soc. géol. France* (7), **5**, 969–979.

Blaison, J. 1968. Affinités, répartition et typologie du genre *Bouleiceras* Thevenin 1906. *Ann. Sci. Univ. Besançon, Ser. 3, Geol.* **5**, 41–49.

Blind, W. 1963. Die Ammoniten des Lias alpha aus Schwaben, von Fonsjoch und Breitenberg (Alpen) und ihre Entwicklung. *Palaeontogr.* **121A**, 38–131.

Blumenthal, M., Durand Delga, M. and Fallot, P. 1958. Données nouvelles sur le tithonique, le crétacé et l'éocène inférieur de la zone marno-schisteuse du Rif septentrional (Maroc). *Notes Mém. Serv. géol. Maroc*, **143**, 35–58.

Bonnefous, J. 1967. Jurassic stratigraphy of Tunisia: a tentative synthesis (northern and central Tunisia, Sahel and Chotts areas). In: L. Martin (editor), *Guidebook to the geology and history of Tunisia*, Petrol. Expl. Soc. Libya, 9th Ann. Field Conf., Breumelhof, Amsterdam.

References

Bonnefous, J. and Rakus, M. 1965. Précisions nouvelles sur le Jurassique du Djebel Bou Kornine d'Hammam-Lif (Tunisie). *Bull. Soc. géol. France*. (7), **7**, 855–859.

Bonte, A., Collin, J. J., Godfriaux, I. and Leroux, B. 1958. Le Bathonien de la région de Marquise. Le Wealden du Boulonnais. *Bull. Carte géol. Fr.* No. 255, **56**, 1–28.

Borchert, H. 1960. Genesis of marine sedimentary iron ores. *Bull. Inst. Min. Metall.* **69**, 261–279.

Borns, H. W. and Hall, B. A. 1969. A reinvestigation of the Mawson Tillite, Victoria Land, Antarctica. *Science* **166**, 870–872.

Bosellini, A. and Broglio-Loriga, C. B. 1971. 'Calcari Grigi' di Rotzo (Giurassico Inferiore, Al topiano d'Asiago). *Ann. Univ. Ferrara* Sez. 9, **5**, 1–61.

Bosellini, A. and Hsü, K. J. 1973. Mediterranean plate tectonics and Triassic palaeogeography. *Nature* **244**, 144–146.

Boswell, P. G. H. 1924. The petrography of the sands of the Upper Lias of Somersetshire. *Geol. Mag.* **61**, 246–264.

Bottke, H., Dengler, H., Finkenwirth, A. *et al.* 1969. Die marin-sedimentaren Eisenerze des Jura in Nordwestdeutschland. *Beih. Geol. Jb.* **79**.

Bouroullec, J. and Deloffre, R. 1969. Interpretation sédimentologique et paléogéographique des microfacies Jurassiques du sud-ouest Aquitain. *Bull. Centre Rech. Pau SNPA* **3**, 287–328.

Bowen, R. 1961. Paleotemperature analyses of Belemnoidea and Jurassic paleoclimatology. *J. Geol.* **69**, 309–320.

Bowen, R. 1966. Oxygen isotopes as climatic indicators. *Earth. Sci. Rev.* **2**, 199–224.

Brand, E. and Hoffmann, K. 1963. Stratigraphy and facies of the north-west German Jurassic and genesis of its oil deposits. *Proc. 6th World Petrol. Congr. Sect. 1*, 223–245.

Bremer, H. 1965. Zur Ammonitenfauna und Stratigraphie des unteren Lias (Sinemurium bis Carixium) in der Umgebung von Ankara (Turkei). *N. Jb. Geol. Paläont. Abh.* **122**, 127–221.

Brenner, R. L. and Davies, D. K. 1973. Storm-generated coquinoid sandstones: genesis of high-energy marine sediments from the Upper Jurassic of Wyoming and Montana. *Bull. geol. Soc. Am.* **84**, 1685–1698.

Brenner, R. L. and Davies, D. K. 1974. Oxfordian sedimentation in Western Interior United States. *Bull. Am. Ass. Petrol. Geol.* **58**, 407–428.

Bretsky, P. W. and Lorenz, D. M. 1971. Adaptive response to environmental stability: a unifying concept in paleoecology. *Proc. N. Am. Paleont. Convent. E.* 522–550. Chicago.

Briden, J. C. 1967. Recurrent continental drift of Gondwanaland. *Nature* **215**, 1334–1339.

Brinkmann, R. 1972. Mesozoic troughs and crustal structure in Anatolia. *Bull. geol. Soc. Am.* **83**, 819–826.

Brockamp B. 1944. Zur paläogeographie und Bitumenfuhrung des Posidonienschiefers im deutschen Lias. *Arch. Lagerstätterforsch.* **77**, 1–59.

Brongersma-Sanders, M. 1971. Origin of major cyclicity of evaporites and bituminous rocks: an actualistic model. *Mar. Geol.* **11**, 123–144.

Brookfield, M. E. 1970. Eustatic changes of sea level and orogeny in the Jurassic. *Tectonophys.* **9**, 347–363.

References

Brookfield, M. E. 1971. An alternative to the 'clastic trap' interpretation of oolitic ironstone facies. *Geol. Mag.* **108**, 137–143.

Brookfield, M. E. 1973. The palaeoenvironment of the Abbotsbury Ironstone (Upper Jurassic) of Dorset. *Palaeontology* **16**, 261–274.

Brookfield, M. E. 1974. Palaeogeography of the Upper Oxfordian and Lower Kimmeridgian (Jurassic) in Britain. *Palaeogeog., Palaeoclimatol., Palaeoecol.* **14**, 137–168.

Brown, G., Catt, J. A. and Weir, A. H. 1969. Zeolites of the clinoptilolite-heulandite type in sediments of south-east England. *Min. Mag.* **37**, 480–488.

Brown, P. R. 1963. Algal limestones and associated sediments in the basal Purbeck of Dorset. *Geol. Mag.* **100**, 565–573.

Brown, P. R. 1964. Petrography and origin of some Upper Jurassic beds from Dorset. *J. sedim. Petrol.* **34**, 254–269.

Brunstrom, R. G. W. and Walmsley, P. J. 1969. Permian evaporites in North Sea Basin. *Bull. Am. Ass. Petrol. Geol.* **53**, 870–883.

Bubenicek, L. 1961. Récherches sur la constitution et la répartition des minerais de fer dans l'Aalenien de Lorraine. *Inst. Rech. Siderurgie, Ser. A* No. 262 and *Sciences de la Terre*, **8**.

Buck, E., Hahn, W. and Schadel, K. 1966. Zur stratigraphie des Bajociums und Bathoniums der Schwäbischen Alb. *Jh. Geol. Landesamt Baden–Württ.* **8**, 23–46.

Bullard, E. C., Everett, J. E. and Smith, A. G. 1965. The fit of the continents around the Atlantic. *Phil. Trans. roy. Soc. A* **258**, 41–51.

Bürgl, H. 1965. El 'Jura-Triasico' de Colombia. *Bol. Geol. Min. Minas Petrol., Serv. Geol. Nac.* **12**, 5–31.

Bürgl, H. 1967. The orogenesis in the Andean System of Colombia. *Tectonophys.* **4**, 429–443.

Busnardo, R., Elmi, S. and Mangold, C. 1965. Ammonites calloviennes de Cabra (Andalousie, Espagne). *Tran. Lab. Geol. Fac. Sci. Lyon*, N.S. No. 11, 49–94.

Busson, G. 1967. Mesozoic of southern Tunisia. In: L. Martin (editor), *Guidebook to the geology and history of Tunisia*, Petrol. Expl. Soc. Libya, 9th Ann. Field Conf. Breumelhof, Amsterdam.

Caillère, S. and Kraut, F. 1954. Les gisements de fer du bassin Lorrain. *Mém. Mus. Hist. nat. Paris (C)* **4**.

Callomon, J. H. 1959. The ammonite zones of the Middle Jurassic beds of East Greenland. *Geol. Mag.* **96**, 505–513.

Callomon, J. H. 1965. Notes on Jurassic stratigraphical nomenclature. 1. Principles of stratigraphic nomenclature. *Rep. Carpath.–Balkan Geol. Ass. 7th Congr.*, Part 2, 81–85.

Callomon, J. H. 1970. Geological maps of Carlsberg Fjord – Fossilbjerget area. *Medd. Grønl.* **168**, No. 4, 1–9.

Callomon, J. H. and Cope, J. C. W. 1971. The stratigraphy and ammonite succession of the Oxford and Kimmeridge Clays in the Warlingham borehole. *Bull. Geol. Surv. G.B.* No. 36, 147–176.

Callomon, J. H., Donovan, D. T. and Trümpy, R. 1972. An annotated map of the Permian and Mesozoic formations of East Greenland. *Medd. Grønl* **168**, No. 3.

Calvert, S. E. 1964. Factors affecting distribution of laminated diatomaceous sediments in Gulf of California. In: *Marine Geology of the Geology of the Gulf of*

References

California – a Symposium, Mem. No. 3, Am. Ass. Petrol. Geol., pp. 311–330.

Cameron, J. B. and Chiu Chung, E. S. 1963. Coal resources, Rosewood–Walloon coalfield. *Queensland geol. Surv. Publ.* 310.

Cantaluppi, G. and Montanari, L. 1968. Carixiano superiore e suo passaggio al Domeriano a N.W. di Arzo (Canton Ticino). *Boll. Soc. Paleont. Ital.* **7**, 1–21.

Cantu Chapa, A. 1963. Etude biostratigraphique des ammonites du centre et de l'est du Mexique. *Mem. Soc. géol. France.* No. 99, N.S. **42**.

Cariou, E. 1966. Les faunes d'ammonites et la sédimentation rythmique dans l'Oxfordien supérieur du seuil du Poitou. *Trav. Inst. Geol. Anthropol. Préhist. Fac. Sci. Poitiers* **7**, 47–67.

Cariou, E. 1973. Ammonites of the Callovian and Oxfordian. In: A. Hallam (editor), *Atlas of Palaeobiogeography*, Elsevier, Amsterdam, 287–295.

Carozzi, A. V., Bouroullec, J., Deloffre, R. and Rumeau, J. L. 1972. Micro-facies du Jurassique d'Aquitaine. *Bull. Centre Rech. Pau SNPA* Special Vol. No. 1.

Carroll, D. 1958. Role of clay minerals in the transportation of iron. *Geochim. Cosmochim. Acta* **14**, 1–27.

Cartagena, R. T. 1965. El Triasico y Jurasico del Departamento de Curepto en la Province de Talca. *An. Fac. Cienc. Fis. Mat. Univ. Chile* **22**, 29–46.

Carter, W. D. 1963. Unconformity marking the Jurassic–Cretaceous boundary in the La Ligua area, Aconcagua province, Chile. *U.S. Geol. Surv. Prof. Paper* 450-E, 61–63.

Casey, R. 1962. The ammonites of the Spilsby sandstone and the Jurassic–Cretaceous boundary. *Proc. Geol. Soc. Lond.* No. 1598, 95–100.

Casey, R. 1963. The dawn of the Cretaceous period in Britain. *Bull. S.E. Union Sci. Soc.* **117**, 1–15.

Casey, R. 1967. The position of the Middle Volgian in the English Jurassic. *Proc. Geol. Soc. Lond.* No. 1640, 128–133.

Casey, R. 1971. Facies, faunas and tectonics in late Jurassic–early Cretaceous Britain. In: F. A. Middlemiss, P. F. Rawson and G. Newall (editors), *Faunal Provinces in space and time*, Seel House Press, Liverpool, pp. 153–168.

Casey, R. 1973. The ammonite succession at the Jurassic–Cretaceous Boundary in eastern England. In: R. Casey and P. F. Rawson (editors), *The Boreal Lower Cretaceous*, Seel House Press, Liverpool, pp. 193–266.

Castellarin, A. 1965. Filoni sedimentari nel Giurese di Lappio (Trentino meridionale). *Giorn. Geol.* **33**, 527–546.

Castellarin, A. 1972. Evoluzione paleotettonica sinsedimentoria del limite tra 'Piattaforma Veneta' e 'Bacino Lombardo' a nord di riva del Garda. *Giorn. Geol.* 2nd Series **38**, fasc. 1, 11–212.

Castellarin, A. and Sartori, R. 1973. Desiccation shrinkage and leaching vugs in the Calcari Grigi Infraliassic tidal flat. *Ecl. geol. Helv.* **66**, 339–343.

Catt, J. A., Gad, M. A., Le Riche, H. H. and Lord, A. R. 1971. Geochemistry, micropalaeontology and origin of the Middle Lias ironstones in northeast Yorkshire. *Chem. Geol.* **8**, 61–76.

Cecioni, G. 1958. Preuves en faveur d'une glaciation neo-Jurassique en Patagonie. *Bull. Soc. Géol. France* (6), **8**, 413–436.

Cecioni, G. O. 1961. Nevaden orogeny in northernmost Chile. *Proc. 9th Pacific Sci. Cong.* **12**, 136–143.

Cecioni, G. and Garcia, F. 1960. Stratigraphy of coastal range in Tarapaca province, Chile. *Bull. Am. Ass. Petrol. Geol.* **44**, 1609–1620.

Cecioni, G. and Westermann, G. E. S. 1968. The Triassic–Jurassic marine transition of coastal central Chile. *Pacif. Geol.* **1**, 45–75.

Charig, A. L. 1973. Jurassic and Cretaceous dinosaurs. In: A. Hallam (editor), *Atlas of Palaeobiogeography*, Elsevier, Amsterdam, pp. 339–352.

Chatalov, G. 1967. Stratigraphy of the Jurassic in the central Strandja Mountain. *Bulgar. Akad. Nauk. Geol. Inst., Tr. Geol. Bulgar. Ser. Struct. Tect.* **16**, 145–166. (In Bulgarian with English summary.)

Choubert, G. and Faure Muret, A. 1962. *Evolution du Domaine Atlasique marocain depuis les temps Paléozoiques. Livre. mém. P. Fallot, Vol. 1* (editor M. Durand Delga), Soc. géol. France, pp. 447–527.

Chowns, T. M. 1966. Depositional environment of the Cleveland Ironstone Series. *Nature* **211**, 1286–1287.

Christ, H. A. 1960. Beiträge zur stratigraphie und Paläontologie des Malm von West-Sizilien. *Schweiz paläont. Abh.* **77**.

Christensen, G. B. 1963. Ostracods from the Purbeck–Wealden beds in Bornholm. *Danm. Geol. Unders.* **2**, No. 86.

Cita, M. B. 1965. Jurassic, Cretaceous and Tertiary microfacies from the southern Alps (Northern Italy). *J. Brill, Leiden.*

Clift, W. O. 1956. Sedimentary history of the Ogaden district, Ethiopia. *Rep. 20th Int. Geol. Cong. Petrol. Gas Sympos.* **1**, 89–112.

Colacicchi, R. and Pialli, G. 1967. Dati a conferma di una lacuna dovuta ad emersione nel Giurese del M. Cucco (Appennino Umbro). *Bull. Soc. Geol. Ital.* **86**, 179–192.

Colacicchi, R. and Pialli, G. 1971. Relationship between some peculiar features of Jurassic sedimentation and paleogeography in the Umbro–Marchigiano basin (central Italy). *Ann. Inst. Geol. Publ. Hungar.* **54**, 195–207.

Colacicchi, R. and Praturlon, A. 1965. Stratigraphical and paleogeographical investigations on the Mesozoic shelf-edge facies in eastern Marsica (central Appennines, Italy). *Geol. Rom.* **4**, 89–118.

Collignon, M. 1957. La partie supérieur du Jurassique au nord de l'Analavelona (Sud Madagascar). *C.R. 3ᵉ Congr. P.I.O.S.A., Tananarive* Sect. C, 73–87.

Collignon, M. 1958–1960. *Atlas des fossiles caracteristiques de Madagascar, Parts 1 to 4, Lias to Tithonian. Serv. géol. Tananarive.*

Collignon, M. 1964 a. Echelle chronostratigraphique proposée pour les domaines Indo-Africano-Malgache (Bathonien moyen à Tithonique). *Publ. Inst. Grand-Ducal, Sect. Sci. Nat. Phys. Math. Luxemburg*, 927–931.

Collignon, M. 1964 b. La serie Dogger–Malm dans la region est d'Ankirihitra (N.W. Madagascar) et ses faunes successives. *C.R. Semaine géol. Com. nat. Malgache Géol.* 43–48.

Collignon, M., Rebilly, G. and Roch, E. 1959. Le Lias et le Jurassique moyen de la region de Kandreho (Madagascar). *Bull. Soc. géol. France* (7) **1**, 132–136.

Colo, G. 1961. Contribution a l'étude du Jurassique du Moyen Atlas septentrional. *Notes Mem. Serv. géol. Maroc.* **139**.

Colom, G. 1967. Sur l'interpretation des sédiments profonds de la zone geosynclinale baleare et subbetique (Espagne). *Palaeogeog, Palaeoclimatol.* **3**, 299–310.

References

Cope, J. C. W. 1967. The palaeontology and stratigraphy of the lower part of the Upper Kimmeridge Clay of Dorset. *Bull. Brit. Mus. (Nat. Hist.), Geol.* **15**.

Corvalan Diaz, J. 1957. Ueber marine sedimente des Tithon und Neocom der Gegend von Santiago. *Geol. Rundsch.* **45**, 919–926.

Corvalan Diaz, J. 1959. El titoniano de rio Lenas, prov. de o'Higgins, con una revision del titoniano y neocomiano de la parte chilena del geosynclinal andino. *Inst. Inves. Geol. Chile* **13** (3).

Couper, R. A. 1958. British Mesozoic microspores and pollen grains. *Palaeontogr.* **103***B*, 75–174.

Cowperthwaite, I. A., Fitch, F. J., Miller, J. A., Mitchell, J. G. and Robertson, R. H. S. 1972. Sedimentation, petrogenesis and radioisotopic age of the Cretaceous Fuller's Earth of southern England. *Clay Mins.* **9**, 309–327.

Cox, K. G. 1970. Tectonics and vulcanism of the Karroo Period and their bearing on the postulated fragmentation of Gondwanaland. In: T. N. Clifford and I. G. Gass (editors), *African magmatism and tectonics*, Oliver & Boyd, Edinburgh, pp. 211–235.

Cox, L. R. 1965. Jurassic Bivalvia and Gastropoda from Tanganyika and Kenya. *Bull. Brit. Mus (Nat. Hist.), Geol.* Suppl. 1.

Craig, G. Y. 1965 (editor). *The Geology of Scotland*. Oliver & Boyd, Edinburgh.

Craig, L. C. *et al.* 1955. Stratigraphy of the Morrison and related formations of the Colorado Plateau region. *U.S. Geol. Surv. Bull.* 1009-E, 125–168.

Crescenti, U. 1971. Biostratigraphic correlations in the Jurassic facies of central Italy by means of the microfossils. *Ann. Inst. Geol. Publ. Hungar.* **54**, 209–213.

Crowell, J. C. 1961. Depositional structures from Jurassic boulder beds, East Sutherland. *Trans. Edinb. Geol. Soc.* **18**, 202–220.

Cruys, H. 1955. Contributions a l'étude géologique des monts du Hodna; 2, La région de Tocqueville et de Bordj R'dir. *Bull. Serv. Carte géol. Alger. B*. N.S. No. 4, 195–326.

Curtis, C. D. and Spears, D. A. 1968. The formation of sedimentary iron minerals. *Econ. Geol.* **63**, 257–270.

Dadlez, R. 1964. Outline of the Lias stratigraphy in western Poland and correlation with the Lias of central Poland. *Kwartal. Geologiczny* **8**, 122–144. (Polish with English summary.)

Dadlez, R., Dayzak-Calikowska, K. and Dembowska, J. 1964. *Geological atlas of Poland – stratigraphic and facial problems. fasc. 9, Jurassic*. Publ. Geol. Inst. Warsaw.

Dahm, H. 1966. Stratigraphie und Paläogeographie im Kantabrischen Jura (Spanien). *Beih. geol. Jb.* **44**, 13–54.

Dalziel, I. W. D. and Cortés, R. 1972. Tectonic style of the southernmost Andes and the Antarctandes. *Rep. 24th Int. geol. Congr.* Sect. 3, 316–327.

Daniel, E. J. 1963. B. Syrie intérieure. *Lex. Strat. Internat.* Vol. 3 (editors, D. T. Donovan and J. E. Hemingway), fasc. 10B.

D'Argenio, B. 1966. Le facies littorali mesozoiche nell'Appennino meridionale. *Boll. Soc. Nat. Napoli* **75**, 497–551.

D'Argenio, B. 1967. Geologia del gruppo del Taburno–Camposauro (Appennino Campano). *Atti Acc. Sci. fis. mat. Soc. Naz. Sci. Lett. Arti. Napoli* Ser. 3, 6.

D'Argenio, B. and Scandone, P. 1971. Jurassic facies pattern in the Southern (Campania–Lucania) Apennines. *Ann. Inst. Geol. Publ. Hungar.* **54**, 383–396.

D'Argenio, B. and Vallario, A. 1967. Sedimentazione ritmica nell'Infralias dell'-Italia meridionale. *Boll. Soc. Nat. Napoli* **76**, 1–7.

Davies, D. K. 1967. Origin of friable sandstone-calcareous sandstone rhythms in the Upper Lias of England. *J. sedim. Petrol.* **37**, 1179–1188.

Davies, D. K. 1969. Shelf sedimentation: an example from the Jurassic of Britain. *J. sedim. Petrol.* **39**, 1344–1370.

Davies, P. J. 1971. Calcite precipitation and recrystallisation fabrics – their significance in Jurassic limestones of Europe. *J. geol. Soc. Austral.* **18**, 279–292.

Davies, R. G. and Gardezi, A. H. 1965. The presence of *Bouleiceras* in Hazara and its geological implications. *Geol. Bull. Punjab. Univ.* No. 5, 1–58.

Dayczak-Calikowska, K. 1967. Problems of Middle Jurassic stratigraphy in Poland. *Biul. Inst. Geol. Wars.* **203**, 72–83.

Dean, W. T., Donovan, D. T. and Howarth, M. K. 1961. The Liassic ammonite zones and subzones of the north-west European Province. *Bull. Brit. Mus. (Nat. Hist.), Geol.* **4**, 438–505.

De Jersey, N. J. 1960. Jurassic spores and pollen grains from the Rosewood coalfield. *Queensland Geol. Surv. Pub.* 294.

De Lapparent, A. F., Blaise J., Lys, M. and Mouterde, R. 1966. Presence du Permien, du Lias et du Jurassique dans la région d'Urasgan (Afghanistan central). *C.R. Acad. Sci. Paris.* **263**, 805–807.

Dercourt, J. 1964. Contribution a l'étude géologique d'un secteur du Peloponnese septentrional. *Ann. Geol. Pays. Hellen.* **15**, 1–408.

Derin, B. and Gerry, E. 1972. *Jurassic biostratigraphy and environments of deposition in Israel.* Israel Inst. Petrol. Rep. 2/72.

Derin, B. and Reiss, Z. 1966. *Jurassic microfacies of Israel.* Israel Inst. Petrol. Spec. Publ.

Desio, A. *et al.* 1963. Stratigraphic studies in the Tripolitanian Jebel (Libya). *Riv. Ital. Paleont. Strat.* **9**, 1–126.

Desio, A., Cita, M. B. and Premoli Silva, I. 1965. The Jurassic Karkar Formation in north-east Afghanistan. *Riv. Ital. Paleont. Strat.* **71**, 1181–1224.

Desio, A., Rossi Ronchetti, C. and Invernizzi, G. 1960. Il giurassico dei dintorni di Jefren in Tripolitaria. *Riv. Ital. Paleont. Strat.* **66**, 65–113.

Dettmann, M. E. 1963. Upper Mesozoic microfloras from south-eastern Australia. *Proc. roy. Soc. Victoria* **77**, 1–148.

Dewey, J. F., Pitman, W. C., Ryan, W. B. F. and Bonnin, J. 1973. Plate tectonics and the evolution of the Alpine system. *Bull. geol. Soc. Am.* **84**, 3134–3180.

Dickinson, W. R. 1962. Petrogenetic significance of geosynclinal andesitic volcanism along the Pacific margin of North America. *Bull. geol. Soc. Am.* **73**, 1241–1256.

Dietz, R. S. and Holden, J. C. 1970. Reconstruction of Pangaea: Breakup and dispersion of continents, Permian to present. *J. geophys. Res.* **75**, 4939–4956.

Dingle, R. V. and Klinger, H. C. 1971. Significance of U. Jurassic sediments in the Knysna Outlier (Cape Province) for timing of the breakup of Gondwanaland. *Nature Phys. Sci.* **232**, 37.

Dingle, R. V. and Klinger, H. C. 1972. The stratigraphy and ostracod fauna of the Upper Jurassic sediments from Breton, in the Knysna Outlier, Cape Province. *Trans. roy. Soc. S. Afr.* **40**, 279–297.

References

Donovan, D. T. 1957. The Jurassic and Cretaceous systems in East Greenland. *Medd. Grønl.* **155**, No. 4, 1–214.

Donovan, D. T. 1958. The ammonite zones of the Toarcian (Ammonitico Rosso facies) of southern Switzerland and Italy. *Ecl. Geol. Helv.* **51**, 33–60.

Donovan, D. T. 1964. Stratigraphy and ammonite fauna of the Volgian and Berriasian ammonite fauna of East Greenland. *Medd. Grønl.* **154**, No. 4, 1–34.

Donovan, D. T. 1967. The geographical distribution of Lower Jurassic ammonites in Europe and adjacent areas. *Systematics Assoc. Publ.* No. 7, 111–132.

Donovan, D. T. and Hemingway, J. E. (editors). 1963. Angleterre, Pays de Galles et Ecosse – Jurassique. *Lex. Strat. Internat.* Vol. 1, fasc. 3ax.

Donze, P. 1958. Les couches de passage du jurassique ou crétacé dans le Jura français et sur les pourtours de la 'fosse vocontienne' (massifs sub-alpins septentrionaux, Ardèche, Grands-Causses, Provence, Alpes-Maritimes). *Trav. Fac. Sci. Lab. Geol. Lyons*, N.S. No. 3.

Donze, P. and Enay, R. 1961. Les cephalopodes du Tithonique inférieur de la Croix de-Saint-Concors près Chambéry (Savoie). *Trav. Lab. Geol. Lyon* N.S. No. 7.

Dubar, G. 1948. La faune domerienne du Jeble bou-Dahar. *Notes Mem. Serv. géol. Maroc.* **68**.

Dubar, G. 1962. Notes sur la paléogeographie du Lias marocain (domaine atlasique). *Livre mém. Prof. P. Fallot* (editor M. Durand Delga), Mem. Soc. géol. France, pp. 529–544.

Dubar, G. 1967. Brachiopodes jurassique du Sahara Tunisien. *Ann. Paleont.* **53**, 33–48.

Dubar, G. and Mouterde, R. 1957. Extension du kimmeridgien marine dans les Asturies depuis Ribadesella jusqu a Gijon. *C.R. Acad. Sci. Paris* **244**, 99–101.

Dubar, G., Peyre, N. and Peyre, Y. 1960. Observations nouvelles sur le Jurassique inférieur et moyen dans les Cordillères bétiques sur la transversale de Malaga (Andalousie, Espagne). *Bull. Soc. géol. France* (7) **2**, 330–339.

Dubar, G., Mouterde, R., Virgili, C. and Suarez, L. C. 1971. El Jurassico de Asturias (norte de Espana). *Cuad. Geol. Iberica* **2**, 561–580.

Dubertret, L. 1963. Liban et Syrie: Chaine des grands massifs éotiers et confins a l'Est. *Lex. Strat. Internat.* Vol. 3, fasc. 10A.

Du Dresnay, R. 1963. Données stratigraphiques complementaires sur le Jurassique moyen des synclinaux d'El Mers et de Skoura (Moyen-Atlas, Maroc). *Bull. Soc. géol. France* (7) **5**, 883–900.

Du Dresnay, R. 1964 a. Les discontinuités de sédimentation pendant le Jurassique, dans la partie orientale du domaine Atlasique marocain, leurs consequences stratigraphiques et leurs relations avec l'orogenese Atlasique. *Publ. Inst. Grand-Ducal Sect. Sci. Nat., Phys. Math. Luxembourg*, 899–912.

Du Dresnay, R. 1964 b. Les découvertes, dans le Dogger du Maroc, de faunes d'ammonites de la province arabique a *Ermoceras*: historique, localisations et répartition paléogéographique. *C.R. Acad. Sci. Paris* **259**, 4754–4757.

Duff, P. McL. D., Hallam, A. and Walton, E. K. 1967. *Cyclic sedimentation*. Elsevier, Amsterdam.

Dunham, K. C. 1960. Syngenetic and diagenetic mineralisation in Yorkshire. *Proc. Yorks. geol. Soc.* **32**, 229–284.

Dunnington, H. V., Wetzel, R. and Morton, D. M. 1959. Mesozoic and Tertiary of Iraq. *Lex. Strat. Internat.* Vol. 3.

Durand Delga, M. and Fallot, P. 1957. Indices de la présence du tithonique et du néocomien dans la dorsale calcaire du Rif (Maroc). *C.R. Acad. Sci. Paris* **245**, 2441–2447.

Dzotsenidze, G. D. 1968. Essay of comparison of the Meso-Cenozoic magmatism of the Caucasus, Crimea, Balkans, and Carpathians. *23rd Internat. Geol. Congr.* Sect. 2, 87–98.

Edmonds, E. A., Poole, E. G. and Wilson, V. 1965. *Geology of the country around Banbury and Edge Hill.* Mem. geol. Surv. G.B.

Einsele, G. and Mosebach, R. 1955. Zur petrographie, Fossilerhaltung und Entstehung der Gesteine des Posidoneinschiefers im Schwäbischen Jura. *N. Jb. Geol. Paläont. Abh.* **101**, 319–430.

Elmi, S. 1967. Le Lias supérieur et le Jurassique moyen de l'Ardèche. *Doc. Lab. Geol. Fac. Sci. Lyon.* No. 19 (fasc. 1 and 2), 1–507.

Elmi, S. 1971. Les faunes a *Prohecticoceras* dui Bathonien inférieur et moyen des confins Algero-Marocains. *Geobios* **4**, 243–263.

Elmi, S. and Mouterde, R. 1965. Le Lias inférieur et moyen entre Aubenas et Privas (Ardèche). *Trav. Lab. Geol. Fac. Sc. Lyon* N.S. No. 12, 143–246.

Emberger, J. 1960. Esquisse géologique de la partie orientale des Mons des Oulad Naïl (Atlas saharien, Algerie). *Publ. Serv. Carte géol. Algerie.* N.S. Bull. 27.

Enay, R. 1966. L'Oxfordien dans la moitié sud du Jura français – étude stratigraphique. *Nouv. Arch. Museum Hist. nat. Lyon.*, fasc. 8, Vols. 1 and 2.

Enay, R. 1972. Paléobiogéographie des ammonites du Jurassique terminal (Tithonique s.l./Volgien s.l./Portlandien s.l.) et mobilité continentale. *Geobios* **5**, 355–407.

Enay, R. 1973. Upper Jurassic (Tithonian) ammonites. In: A. Hallam (editor), *Atlas of Palaeobiogeography*, Elsevier, Amsterdam, pp. 297–308.

Enay, R., Martin, C., Monod, O. and Thieuloy, J. P. 1971. Jurassic supérieur à ammonites (Kimmeridgien–Tithonique) dans l'autochtone du Taurus de Beysehir (Turquie meridonale). *Ann. Inst. Geol. Publ. Hungar.* **54**, 397–422.

Epstein, S. and Mayeda, T. 1953. Variations in O^{18} content of waters from natural sources. *Geochim. Cosmochim. Acta* **27**, 213–224.

Erben, H. K. 1956a. El Jurassico inferior de Mexico y sus amonitas. *20th Int. geol. Congr.*, Mexico.

Erben, H. K. 1956b. El Jurassico medio y el Calloviano de Mexico. *20th Int. Geol. Congr.*, Mexico.

Erben, H. K. 1957a. Paleogeographic reconstructions for the Lower and Middle Jurassic and for the Callovian of Mexico. *Rep. 20th Int. Geol. Congr.* Sect. 11, 35–40.

Erben, H. K. 1957b. New biostratigraphic correlations in the Jurassic of eastern and south-central Mexico. *Rep. 20th Int. Geol. Congr.* Sect. 11, 43–52.

Ernst, W. 1967. Die Liastongrube Grimmen: Sediment, Makrofauna und Stratigraphie – ein Uberblick. *Geologie* **5**, 550–569.

Evernden, J. F. and Richards, J. R. 1962. Potassium–argon ages in eastern Australia. *J. geol. Soc. Australia* **9**, 1–50.

Fabricius, F. H. 1966. *Beckensedimentation und Riffbildung an der Wende Trias/Jura in den Bayerisch–Tiroler Kalkalpen.* Brill, Leiden.

References

Fabricius, F. 1967. Die Rät- und Lias-Oolithe der nordwestlichen Kalkalpen. *Geol. Rundsch.* **56**, 180–190.

Fabricius, F., Friedrichsen, V. and Jacobshagen, V. 1970. Paläotemperaturen und Paläoklima in Obertrias und Lias der Alpen. *Geol. Rundsch.* **59**, 805–826.

Falcon, N. L. and Kent, P. E. 1960. Geological results of petroleum exploration in Britain 1945–1957. *Mem. Geol. Soc. Lond.* No. 2.

Fantini Sestini, N. 1962. Contributo allo studio della ammoniti del Domeriano di Monte Domaro (Brescia). *Riv. Ital. Paleont. Stratig.* **68**, 483–554.

Fantini Sestini, N. 1966. Upper Liassic molluscs from Shemshak Formation. *Inst. Paleont. Univ. Milano* Ser. P., Pub. 154, 795–852.

Farag, I. A. M. 1959. Contribution to the study of the Jurassic formations in the Maghara massif (northern Sinai). *Egypt. J. Geol.* **3**, 175–199.

Farinacci, A. 1964. Micro-organismi dei Calcari 'Maiolica' e 'Scaglia' osservati al Microscopio Elletronico (Nannoconi et Coccolithophoridi). *Boll. Soc. Paleont. Ital.* **3**, 172–181.

Farinacci, A. 1967. La serie giurassica–neocomiana di Monte Lacerone (Sabina). Nuove vedute sull' interpretazione paleogeografica delle aree di facies umbro–marchigiana. *Geologia romana* **6**, 421–480.

Farrow, G. E. 1966. Bathymetric zonation of Jurassic trace fossils from the coast of Yorkshire, England. *Palaeogeog. Palaeoclimatol. Palaeoecol.* **2**, 103–151.

Fatmi, A. N. 1972. Stratigraphy of the Jurassic and Lower Cretaceous rocks and Jurassic ammonites from northern areas of west Pakistan. *Bull. Brit. Mus. (Nat. Hist.), Geol.* **20**, 302–380.

Faul, H., Stern, T. W., Thomas, H. H. and Elmore, P. L. D. 1963. Ages of intrusion and metamorphism in the northern Appalachians. *Am. J. Sci.* **261**, 1–19.

Fenninger, A. 1967. Riffentwicklung im oberostalpinen Malm. *Geol. Rundsch.* **56**, 171–185.

Fenninger, A. and Holzer, H. L. 1970. Fazies und Paläogeographie des oberostalpinen, *Malm. Mitt. Geol. Ges. Wien* **63**, 52–141.

Ferguson, H. G. and Muller, S. W. 1949. Structural geology of the Hawthorne and Tonopah quadrangles, Nevada. *U.S. Geol. Surv. Prof. Paper* 216.

Fischer, A. G. 1964. The Lofer Cyclotherms of the Alpine Triassic. *Bull. geol. Surv. Kansas* **169**, 107–149.

Fischer, A. G. and Garrison, R. E. 1967. Carbonate lithification on the sea floor. *J. Geol.* **75**, 488–496.

Fischer, J. C. 1969. Géologie, paléontologie et paléoecologie du Bathonien au Sud-ouest du Massif Ardennais. *Mem. Mus. Nat. Hist. Nat. N.S.* Ser. C, **20**.

Fischer, R. 1966. Die Dactylioceratidae (Ammonoidea) der Kammerker (Nordtirol) und die zonengliederung der alpinen Toarcien. *Bayer. Akad. Wiss. Math.–Naturwiss. Kl.*, Abh. N.S. **126**, 1–83.

Fleming, C. A. 1962. New Zealand biogeography: a palaeontologist's approach. *Tuatara* **20**, 53–108.

Fleming, C. A. 1970. The Mesozoic of New Zealand: chapters in the history of the Circum-Pacific mobile belt. *Quart. J. Geol. Soc. Lond.* **125**, 126–170.

Florin, R. 1963. The distribution of conifer and taxad genera in time and space. *Acta Hortic. Bergen.* **20**, 121–312.

Flügel, E. 1967. Elektronenmikroskopische Untersuchungen an mikritischen Kalken. *Geol. Rundsch.* **56**, 341–358.

References

Flügel, H. W. and Fenninger, A. 1966. Die Lithogenese der Oberalmer Schichten und der mikritischen Plassen-Kalke (Tithonium, Nordliche Kalkalpen). *N. Jb. Geol. Paläont. Abh.* **123**, 249–280.

Foland, K. A., Quinn, A. W. and Giletti, B. J. 1971. K–Ar and Rb–Sr Jurassic and Cretaceous ages for intrusives of the White Mountain magma series northern New England. *Am. J. Sci.* **270**, 321–330.

Frebold, H. 1960. The Jurassic faunas of the Canadian Arctic, Lower Jurassic and lowermost Middle Jurassic ammonites. *Bull. Geol. Surv. Canada* **59**.

Frebold, H. 1961. The Jurassic faunas of the Canadian Arctic – Middle and Upper Jurassic ammonites. *Bull. Geol. Surv. Canada* **74**.

Frebold, H. 1964. The Jurassic faunas of the Canadian Arctic – Caloceratinae. *Bull. Geol. Surv. Canada* **119**.

Frebold, H. and Tipper, H. W. 1970. Status of the Jurassic in the Canadian Cordillera of British Columbia, Alberta and southern Yukon. *Can. J. Earth Sci.* **7**, 1–21.

Frebold, H. and Tipper, H. W. 1973. Upper Bajocian–Lower Bathonian ammonite fauna and stratigraphy of Smithers area, British Columbia. *Can. J. Earth Sci.* **10**, 1109–1131.

Freeland, G. L. and Dietz, R. S. 1971. Plate tectonic evolution of Caribbean– Gulf of Mexico region. *Nature* **232**, 20–23.

Friedman, G. M., Barzel, A. and Derin, B. 1971. Paleoenvironments of the Jurassic in the coastal belt of northern and central Israel and their significance in the search for petroleum reservoirs. *Geol. Surv. Israel Rep.* OD/1/71, 1–26.

Fülöp, J. 1971. Les formations jurassiques de la Hongrie. *Ann. Inst. Geol. Publ. Hungar.* **54**, 31–46.

Fürsich, F. T. 1971. Hartgründe und Kondensation im Dogger von Calvados. *N. Jb. Geol. Paläont. Ahb.* **138**, 313–342.

Fürsich, F. T. 1973. *Thalassinoides* and the origin of nodular limestones in the Corallian Beds (Upper Jurassic) of southern England. *N. Jb. Geol. Paläont. Mh.* **3**, 136–156.

Fürsich, F. T. 1974. Corallian trace fossils from England and Normandy. *Stuttgarter Beitr. Naturk.*, Ser. B, No. 4, 1–52.

Gabilly, J. 1964 a. Le Jurassique inférieur et moyen sur le littoral vendéen. *Trav. Inst. Géol. Anthropol. Prehist. Fac. Sci. Poitiers* **5**, 69–107.

Gabilly, J. 1964 b. Stratigraphie et limites de l'étage Toarcien à Thouars et dans les régions voisines. *Publ. Inst. Grand-Ducal, Sect. Sci. Nat. Phys. Math. Luxembourg*, 193–201.

Gad, M. A., Catt, J. A. and Le Riche, H. H. 1969. Geochemistry of the Whitbian (Upper Lias) sediments of the Yorkshire coast. *Proc. Yorks. geol. Soc.* **37**, 105–136.

Galacz, A. and Vörös, A. 1972. Jurassic history of the Bakony Mountains and interpretation of principal lithological phenomena. *Bull. Hungar. Geol. Soc.* **102**, 122–135. (In Hungarian, with English summary.)

Garrison, R. E. 1967. Pelagic limestones of the Oberalm Beds (Upper Jurassic– Lower Cretaceous), Austrian Alps. *Bull. Canad. Petrol. Geol.* **15**, 21–49.

Garrison, R. E. and Fischer, A. G. 1969. Deep water limestones and radiolarites of the Alpine Jurassic. *Soc. Econ. Paleont. Min. Spec. Publ.* **14**, 20–55.

References

Gatrall, M., Jenkyns, H. C. and Parsons, C. F. 1972. Limonitic concretions from the European Jurassic, with particular reference to the 'snuff-boxes' of southern England. *Sedimentol.* **18**, 79–103.

Gebelein, C. D. 1969. Distribution, morphology and accretion rate of Recent subtidal algal stromatolites. *J. sedim. Petrol.* **39**, 49–69.

Geczy, B. 1961. Die Jurassiche schichtreihe des tuzkoves–Grabens von Bakonyc-sernye. *Ann. Inst. Geol. Publ. Hungar.* **49**, 507–563.

Geczy, B. 1966. Ammonoides jurassique de Czernye, Montagne Bakony, Hongrie. *Geol. Hungarica Ser. Palaeont.* fascs. 34 and 35.

Geel, T. 1966. Biostratigraphy of Upper Jurassic and Cretaceous sediments near Caravaca (S.E. Spain) with special emphasis on *Tintinnina* and *Nannoconus*. *Geol. Mijnb.* **45**, 375–385.

George, T. N. *et al.* 1969. Recommendations on stratigraphical usage. *Proc. geol. Soc. Lond.* No. 1656, 139–166.

Gerasimov, P. A. and Mikhailov, N. P. 1966. The Volgian stage and the standard scale for the upper series of the Jurassic System. *Isv. Akad. Nauk. SSSR, Ser. Geol.* No. 2, 118–138. (In Russian.)

Gerth, H. 1955. *Der geologische Bau der Südamerikanischen Kordillere.* Born-traeger, Berlin.

Gerth, H. 1965. Ammoniten des mittleren und oberen Jura und der älteste kreide von Nordabhang des Schneegebirges in Neu Guinea. *N. Jb. Geol. Paläont. Abh.* **121**, 209–218.

Gervasio, F. C. 1967. Age and nature of orogenesis of the Philippines. *Tectono-physics* **4**, 367–378.

Geyer, O. F. 1961. Monograph der Perisphinctidae des Unteren Unterkimerid-gium (Weisser Jura, Badenerschichten) im süddeutschen Jura. *Palaeontogr.* **117A**, 1–157.

Geyer, O. F. 1967a. Zur faziellen Entwicklung des subbetischen Juras in Sud-spanien. *Geol. Rundsch.* **56**, 973–992.

Geyer, O. F. 1967b. Das Typus-Profil der Morrocoyal Formation (Unterlias; Dept. Bolivar, Kolumbien). *Mitt. Inst. Colombo–Aleman Invest. Cient.* **1**, 53–63.

Geyer, O. F. 1968. Über den Jura der Halbinsel La Guajira (Kolumbien). *Mitt. Inst. Colombo–Aleman Invest. Cient.* **2**, 67–83.

Geyer, O. F. 1973. Das präkretazische Mesozoikum von Kolombien *Geol. Jb.*, Ser. B, **5**, 1–156.

Geyer, O. F. and Gwinner, M. P. 1968. *Einführung in die Geologie von Baden–Württemberg.* Schweizerb., Stuttgart.

Geyer, O. F. and Hinkelbein, K. 1971. Eisenoolithische Kondensations–Horizonte im Lias der Sierra de Espuña (Provinz Murcia, Spanien). *N. Jb. Geol. Paläont. Mh.* **7**, 398–414.

Gilluly, J. 1963. The tectonic evolution of the western United States. *Quart. J. geol. Soc. Lond.* **119**, 133–174.

Goldberg, M. and Friedman, G. M. In press. Paleoenvironments and paleogeo-graphic evolution of the Jurassic System in southern Israel. *Bull. geol. Surv. Israel.*

Gordon, W. A. 1970. Biogeography of Jurassic Foraminifera. *Bull. geol. Soc. Am.* **81**, 1689–1704.

Gorsky, I. I. and Leonenok, N. L. 1964. Stratigraphie des sédiments continentaux jurassiques de l'U.R.S.S. *Publ. Inst. Grand-Ducal Sect. Sci. Nat., Phys., Math. Luxembourg*, 721–745.

Grasmück, K. 1961. Die helvetischen sedimente am Nordostrand des Mont Blanc-Massivs (zwischen Sembrancher und dem Col Ferret). *Ecl. Geol. Helv.* **54**, 353–450.

Green, A. G. 1972. Sea floor spreading in the Mozambique Channel. *Nature Phys. Sci.* **236**, 19–21.

Green, G. W. and Donovan, D. T. 1969. The Great Oolite of the Bath area. *Bull. geol. Surv. G.B.* No. 30, 1–63.

Green, G. W. and Melville, R. V. 1956. The stratigraphy of the Stowell Park borehole. *Bull. geol. Surv. G.B.* No. 11, 1–66.

Green, G. W. and Welch, F. B. A. 1965. *Geology of the country around Wells and Cheddar*. Mem. geol. Surv. G.B.

Greenwood, J. E. G. W. and Bleackley, D. 1967. Geology of the Arabian Peninsula: Aden Protectorate. *U.S. Geol. Surv. Prof. Paper* 560-C.

Griffin, J. J., Windom, H. and Goldberg, E. D. 1968. Distribution of clay minerals in the world ocean. *Deep-Sea Res.* **15**, 433–461.

Griffon, J-Cl. and Mouterde, R. 1964. Découverte de faunes hettangiennes au S de Tetouan (Rif septentrional, Maroc). *C.R. Soc. géol. France* (7) **6**, 61–63.

Gross, A. 1965. Contribution a l'étude du Jurassique moyen et supérieur des Pre alpes medianes vaudoises. *Ecl. Geol. Helv.* **58**, 743–788.

Grubić, A. 1964. Les bauxites de la province dinarique (Yougoslavie). *Bull. Soc. géol. France* (7) **6**, 382–388.

Grunau, H. R. 1965. Radiolarian cherts and associated rocks in space and time. *Ecl. Geol. Helv.* **58**, 157–208.

Gry, H. 1960. *Geology of Bornholm, Guide to excursion Nos. A45 and C40.* 21st Int. Geol. Congr. Copenhagen.

Guex, J. 1972. Répartition biostratigraphique des ammonites du Toarcien moyen de la bordure sud des Causses (France) et révision des ammonites décrites et figurées par Monestier (1931). *Ecl. geol. Helv.* **65**, 611–644.

Guex, J. 1973. Aperçu biostratigraphique sur le Toarcien inférieur du Moyen-Atlas marocain et discussion sur la zonation de ce sous-étage dans les séries meditérranéennes. *Ecl. geol. Helv.* **66**, 493–523.

Gunn, B. M. and Warren, G. 1962. Geology of Victoria Land between the Mawson and Mullock Glaciers, Ross Dependency, Antarctica. *N.Z. Geol. Surv. Bull.* N.S. No. 71.

Guppy, D. J. *et al.* 1958. The Geology of the Fitzroy Basin, Western Australia, *Bur. Min. Res. Austral., Geol. and Geophys. B.* No. 36.

Gusić, I., Nikler, L. and Sokaćd, B. 1971. The Jurassic in the Dinaric Mountains of Croatia and the problems of its subdivision. *Ann. Inst. Geol. Publ. Hungar.* **54**, 165–184.

Gussow, W. C. 1960. Jurassic–Cretaceous boundary in Western Canada and late Jurassic age of the Kootenay formation. *Trans. roy. Soc. Canada* Sect. 4, **54**, 45–64.

Gygi, R. 1966. Über das zeitliche Verhältnis zwischen der transversarium-Zone in der Schweiz und der plicatilis-Zone in England. *Ecl. geol. Helv.* **59**, 935–942.

References

Gygi, R. 1969. Zur Stratigraphie der Oxford–Stufe der Nordschweiz und des südeutschen Grenzgebietes. *Beitr. geol. Karte Schweiz*, N.S. **136**.

Haas, O. 1955. Revision of the Jurassic ammonite fauna of Mount Hermon, Syria. *Bull. Am. Mus. Nat. Hist.* **108**.

Hales, A. L. 1960. Research at the Bernard Price Institute of Geophysical Research, University of Witwatersrand, Johannesburg. *Proc. roy. Soc. A* **258**, 1–26.

Hallam, A. 1958. The concept of Jurassic axes of uplift. *Science Progress* No. 183, 441–449.

Hallam, A. 1960. A sedimentary and faunal study of the Blue Lias of Dorset and Glamorgan. *Phil. Trans. roy. Soc. B* **243**, 1–44.

Hallam, A. 1961. Cyclothems, trangressions and faunal change in the Lias of North West Europe. *Trans. Edinb. Geol. Soc.* **18**, 132–174.

Hallam, A. 1962. A band of extraordinary calcareous concretions in the Upper Lias of Yorkshire, England. *J. sed. Petrol.* **32**, 840–47.

Hallam, A. 1963 a. Observations on the palaeoecology and ammonite sequence of the Frodingham Ironstone (Lower Jurassic). *Palaeontology* **6**, 554–574.

Hallam, A. 1963 b. Eustatic control of major cyclic changes in Jurassic sedimentation. *Geol. Mag.* **100**, 444–450.

Hallam, A. 1964. Origin of the limestone–shale rhythm in the Blue Lias of England: a composite theory. *J. Geol.* **72**, 157–169.

Hallam, A. 1965. Observations on marine Lower Jurassic stratigraphy of North America, with special reference to United States. *Bull. Am. Ass. Petrol. Geol.* **48**, 1485–1501.

Hallam, A. 1966. Depositional environment of British Liassic ironstones considered in the context of their facies relationships. *Nature* **209**, 1306–1307.

Hallam, A. 1967 a. Siderite- and calcite-bearing concretionary nodules in the Lias of Yorkshire. *Geol. Mag.* **104**, 222–227.

Hallam, A. 1967 b. Sedimentology and palaeogeographic significance of certain red limestones and associated beds in the Lias of the Alpine region. *Scot. J. Geol.* **3**, 195–220.

Hallam, A. 1967 c. An environmental study of the Upper Domerian and Lower Toarcian in Great Britain. *Phil. Trans. roy. Soc. B* **252**, 393–445.

Hallam, A. 1969 a. Faunal realms and facies in the Jurassic. *Palaeontology* **12**, 1–18.

Hallam, A. 1969 b. Tectonism and eustasy in the Jurassic. *Earth Sci. Rev.* **5**, 45–68.

Hallam, A. 1969 c. A pyritised limestone hardground in the Lower Jurassic of Dorset (England). *Sedimentology* **12**, 231–240.

Hallam, A. 1970. *Gyrochorte* and other trace fossils in the Forest Marble (Bathonian) of Dorset, England. In: T. P. Crimes and J. C. Harper (editors), *Trace fossils*, Seel House Press, Liverpool, pp. 89–100.

Hallam, A. 1971 a. Mesozoic geology and the opening of the North Atlantic. *J. Geol.* **79**, 129–157.

Hallam, A. 1971 b. Provinciality in Jurassic faunas in relation to facies and palaeogeography. In: F. A. Middlemiss, P. F. Rawson and G. Newall (editors), *Faunal Provinces in space and time*, Liverpool, Seel House Press, pp. 129–152.

Hallam, A. 1971 c. Facies analysis of the Lias in West Central Portugal. *N. Jb. Geol. Paläont. Abh.* **139**, 226–265.

Hallam, A. 1972 a. Diversity and density characteristics of Pliensbachian–Toarcian molluscan and brachiopod faunas of the North Atlantic. *Lethaia* 5, 389–412.

Hallam, A. 1972 b. Relation of Palaeogene ridge and basin structures and vulcanicity in the Hebrides and Irish Sea regions of the British Isles to the opening of the North Atlantic. *Earth Planet, Sci. Lett.* 16, 171–177.

Hallam, A. 1973 (editor). *Atlas of Palaeobiogeography*, Elsevier, Amsterdam.

Hallam, A. In press. Coral patch reefs in the Bajocian (Middle Jurassic) of Lorraine. *Geol. Mag.*

Hallam, A. and Payne, K. W. 1958. Germanium enrichment in lignites from the Lower Lias of Dorset. *Nature* 181, 1008–1009.

Hallam, A. and Sellwood, B. W. 1968. Origin of Fuller's Earth in the Mesozoic of southern England. *Nature* 220, 1193–1195.

Hallam, A. and Sellwood, B. W. 1970. Montmorillonite and zeolites in Mesozoic and Tertiary beds of southern England. *Min. Mag.* 37, 950–952.

Häntzschel, W. and Reineck, H. E. 1968. Fazies-Untersuchurgen im Hettangium von Helmstedt (Niedersachsen). *Mitt. geol. Staatsinst. Hamburg* 37, 5–39.

Harland, W. B. 1971. Introduction. In: *The Phanerozoic time scale: a supplement*. Geol. Soc. London Spec. Publ. No. 5, 3–7.

Harland, W. B. *et al.* 1972. A concise guide to stratigraphical procedure. *J. geol. Soc.* 128, 295–305.

Harrington, H. J. 1961. Geology of parts of Antofagasta and Atacama provinces, northern Chile. *Bull. Am. Ass. Petrol. Geol.* 45, 169–197.

Harrington, H. J. 1965. Geology and morphology of Antarctica. In: P. van Oye and J. van Mieghan (editors), *Biogeography and Ecology in Antarctica*. North-Holland, Amsterdam, pp. 1–71.

Harshbarger, J. W., Repenning, C. A. and Irwin, J. H. 1957. Stratigraphy of the uppermost Triassic and the Jurassic rocks of the Navajo country. *U.S. Geol. Surv. Prof.* Paper 291.

Hauff, B. 1953. *Das Holzmadenbuch*. Rau, Öhringen.

Haug, E. 1907. *Traité de Geologie*. Paris.

Hayami, I. 1961. On the Jurassic pelecypod faunas of Japan. *J. Fac. Sci. Univ. Tokyo* Sect. 2, 243–343.

Hayami, I. 1972. Lower Jurassic Bivalvia from the environs of Saigon. *Geol. Palaeont. S.E. Asia* 10, 179–230.

Hays, J. D. and Pitman, W. C. 1973. Lithospheric plate motion, sea level changes and climatic and ecological consequences. *Nature* 246, 18–22.

Heirtzler, J. R. *et al.* 1973. Age of the floor of the eastern Indian Ocean. *Science* 180, 952–954.

Hemingway, J. E. 1951. Cyclic sedimentation and the deposition of ironstone in the Yorkshire Lias. *Proc. Yorks. geol. Soc.* 28, 67–74.

Hemingway, J. E. 1958. The geology of the Whitby area. In: G. H. J. Daysh (editor), *A survey of Whitby and the surrounding area*, Shakespeare Head Press, Windsor.

Hemingway, J. E. 1974. Jurassic. In: D. H. Rayner and J. E. Hemingway (editors), *The Geology and Mineral Deposits of Yorkshire*, Yorkshire Geol. Soc., pp. 161–223.

Herrmann, A. 1971. Die Asphaltkalk–Lagerstätte bei Holzen/Ith auf der Sudwest flanke der Hils–Mulde. *Beih. Geol. Jb.* 95.

References

Herron, E. M., Dewey, J. F. and Pitman, W. C. 1974. Plate tectonics model for the evolution of the Arctic, *Geology*, August, 377–380.

Hillebrandt, A. von, 1970. Zur Biostratigraphie und Ammoniten-Fauna des südamerikanischen Jura (insbes. Chile). *N. Jb. Geol. Paläont. Abh.* **136**, 166–211.

Hillebrandt, A. von, 1973 a. Die Ammonitengattungen *Bouleiceras* und *Frechiella* im Jura von Chile und Argentinien. *Ecl. geol. Helv.* **66**, 351–363.

Hillebrandt, A. von, 1973 b. Neue Ergebnisse über den Jura in Chile und Argentinien. *Münster Forsch. Geol. Paläont.* **31/32**, 167–199.

Hiller, K. 1964. Über die Bank – und Schwammfazies des Weissen Jura der Schwäbischen Alb (Württemburg). *Arb. Geol. Paläont. Inst. TH Stuttgart* N.S. **40**, 1–190.

Hirano, H. 1973. Biostratigraphic study of the Jurassic Toyora Group. *Trans. Proc. Paleont. Soc. Japan*, N.S. No. 89, 1–14; No. 90, 45–71.

Ho, C. and Coleman, J. M. 1969. Consolidation and cementation of Recent sediments in the Atchafalaya Basin. *Bull. geol. Soc. Am.* **80**, 183–192.

Hoffmann, K. 1966. Die Stratigraphie und Paläogeographie der bituminosen fazies des nordwestdeutschen Oberlias (Toarcium). *Beih. Geol. Jb.* **58**, 443–498.

Hölder, H. 1964. *Jura*. In the series: F. Lotze (editor), *Handbuch der stratigraphischen Geologie*, Enke, Stuttgart.

Hölder, H. and Hollmann, R. 1969. Bohrgänge mariner Organismen in jurassischen Hart – und Felsboden. *N. Jb. Geol. Paläont. Abh.* **133**, 79–88.

Hölder, H. and Ziegler, B. 1959. Stratigraphische und faunistische Beziehungen im Weissen Jura (Kimeridgien) zwischen Süddeutschland und Ardèche. *N. Jb. Geol. Paläont. Abh.* **108**, 150–214.

Hollmann, R. 1964. Subsolutions-Fragmente (Zur Biostratinomie der Ammonoidea in Malm des Monte Baldo/Norditalien). *N. Jb. Geol. Paläont. Abh.* **119**, 22–82.

Hottinger, L. 1971. Larger foraminifera of the Mediterranean Jurassic and their stratigraphic use. *Ann. Inst. Geol. Publ. Hungar.* **54**, 499–504.

Hourcq, V. and Reyre, D. 1956. Les récherches petrolières dans la zone cotière du Gabon (Afrique equatoriale française). *20th Int. Geol. Congr. Mexico. Petrol. Gas Symp.* **1**, 113–141.

Howarth, M. K. 1957. The Middle Lias of the Dorset coast. *Quart. J. Geol. Soc. Lond.* **113**, 185–203.

Howarth, M. K. 1958. Upper Jurassic and Cretaceous ammonite faunas of Alexander Land and Graham Land. *Falk. Is. Dep. Surv. Sci. Rep.* 21.

Howarth, M. K. 1962. The Jet Rock Series and the Alum Shale Series of the Yorkshire coast. *Proc. Yorks. geol. Soc.* **33**, 381–422.

Howarth, M. K. 1964. The Jurassic period. *Quart. J. geol. Soc. Lond.* **120**, Suppl. 203–205. (The Phanerozoic Time Scale.)

Howarth, M. K. 1973. The stratigraphy and ammonite fauna of the Upper Liassic Grey Shales of the Yorkshire coast. *Bull. Brit. Mus. (Nat. Hist.)*, *Geology* 24, 238–277.

Howitt, F. 1964. Stratigraphy and structure of the Purbeck inliers of Sussex (England). *Quart. J. geol. Soc. Lond.* **120**, 77–113.

Howitt, F. 1974. North Sea Oil in a world context. *Nature* **249**, 700–703.

Hoyer, P. 1965. Fazies, Paläogeographie und Tektonik des Malm im Deister, Osterwald und Suntel. *Beih. Geol. Jb.* 61.

References

Hsü, K. J. 1972. Origin of saline giants; a critical review after the discovery of the Mediterranean evaporite. *Earth Sci. Rev.* **8**, 371–396.

Hsü, K. J. and Ohrbom, R. 1969. Mélanges of San Francisco Peninsula – geologic reinterpretation of type Franciscan. *Bull. Am. Ass. Petrol. Geol.* **53**, 1348–1367.

Huber, N. K. and Garrels, R. M. 1953. Relation of pH and oxidation potential to sedimentary iron mineral formation. *Econ. Geol.* **48**, 337–357.

Huckriede, R. 1967. Molluskenfanuen mit limnischen und brackischen Elementen aus Jura, Serpulit und Wealden NW-Deutschlands und ihre paläogeographische Bedeutung. *Beih. geol. Jb.* **67**.

Huckriede, R., Kursten, M. and Venzlaff, H. 1962. Zur geologie des Gebietes zwischen Kerman und Sagand. *Beih. Geol. Jb.* **51**.

Hudson, J. D. 1963. The recognition of salinity-controlled mollusc assemblages in the Great Estuarine Series (Middle Jurassic) of the Inner Hebrides. *Palaeontology* **6**, 318–326.

Hudson, J. D. 1964. The petrology of the sandstones of the Great Estuarine Series, and the Jurassic palaeogeography of Scotland. *Proc. Geol. Ass. Lond.* **75**, 499–528.

Hudson, J. D. 1967. Speculations on the depth relations of calcium carbonate solution in Recent and ancient seas. *Mar. Geol.* **5**, 473–480.

Hudson, J. D. 1968. The microstructure and mineralogy of the shell of a Jurassic mytilid (Bivalvia). *Palaeontology* **11**, 163–182.

Hudson, J. D. 1970. Algal limestones with pseudomorphs after gypsum from the Middle Jurassic of Scotland. *Lethaia* **3**, 11–40.

Hudson, J. D. and Jenkyns, H. C. 1969. Conglomerates in the Adnet Limestones of Adnet (Austria) and the origin of the 'Scheck'. *N. Jb. Geol. Paläont. Mh.* **9**, 552–558.

Hudson, J. D. and Palframan, D. F. B. 1969. The ecology and preservation of the Oxford Clay fauna at Woodham, Buckinghamshire. *Quart. J. geol. Soc. Lond.* **124**, 387–418.

Hudson, R. G. S. 1958. The Upper Jurassic faunas of southern Israel. *Geol. Mag.* **95**, 415–425.

Hudson, R. G. S. and Chatton, M. 1959. The Musandam Limestone (Jurassic to Lower Cretaceous) of Oman, Arabia. *Notes Mém. Moy-Orient.* **7**, 69–93.

Hülsemann, J. and Emery, K. O. 1961. Stratification in Recent sediments of Santa Barbara Basin as controlled by organisms and water character. *J. Geol.* **69**, 279–290.

Hyndman, D. W. 1968. Mid-Mesozoic multiphase folding along the border of the Shuswap metamorphic complex. *BGSA* **79**, 575–588.

Imlay, R. W. 1956. Marine Jurassic exposed in Bighorn Basin, Pryor and Bighorn Mountains. *Bull. Am. Ass. Petrol. Geol.* **40**, 562–599.

Imlay, R. W. 1957. Palaeoecology of Jurassic seas in the western interior of the United States. *Mem. geol. Soc. Am.* No. 67, Vol. 2, 469–504.

Imlay, R. W. 1962. Jurassic (Bathonian or early Callovian) ammonites from Alaska and Montana. *U.S. Geol. Surv. Prof. Paper* 374-C.

Imlay, R. W. 1964. Middle Bajocian ammonites from the Cook Inlet region, Alaska. *U.S. Geol. Surv. Prof. Paper* 418–B.

Imlay, R. W. 1965. Jurassic marine faunal differentiation in North America. *J. Paleont.* **39**, 1023–1038.

References

Imlay, R. W. 1967 a. Twin Creek Limestone (Jurassic) in the Western Interior of the United States. *U.S. Geol. Surv. Prof. Paper* 540.

Imlay, R. W. 1967 b. The Mesozoic pelecypods *Otapiria* Marmick and *Lupherella* Imlay, new genus, in the United States. *U.S. Geol. Surv. Prof. Paper* 573-B.

Imlay, R. W. 1970. Some Jurassic ammonites from central Saudi Arabia. *U.S. Geol. Surv. Prof. Paper* 643-D.

Imlay, R. W. and Detterman, R. L. 1973. Jurassic paleobiogeography of Alaska. *U.S. Geol. Surv. Prof. Paper* 801.

Imlay, R. W., Dole, H. M., Wells, F. G. and Peck, D. 1959. Relations of certain Upper Jurassic and Lower Cretaceous formations in southwestern Oregon. *Bull. Am. Ass. Petrol. Geol.* **43**, 2770–2785.

Irwin, M. L. 1965. General theory of epeiric clear water sedimentation. *Bull. Am. Ass. Petrol. Geol.* **49**, 445–459.

Irwin, W. P. 1964. Late Mesozoic orogenies in the ultramafic belts of northwestern California and southwestern Oregon. *U.S. Geol. Surv. Prof. Paper* 501-C.

Jacobshagen, V. 1965. Die Allgäu-Schichten (Jura-Fleckenmergel) zwischen Wettersteingebirge und Rhein. *Jb. geol. B-A. Wien* **108**, 1–114.

Jacobson, R. R. E., Snelling, N. J. and Truswell, J. F. 1964. Age determinations in the geology of Nigeria, with special reference to the Older and Younger Granites. *Overseas Geol. Min. Res.* **9**, 168–182.

James, G. A. and Wynd, J. G. 1965. Stratigraphic nomenclature of Iranian Oil Consortium Agreement area. *Bull. Am. Ass. Petrol. Geol.* **49**, 2182–2245.

Jansa, L. 1972. Depositional history of the coal-bearing Upper Jurassic–Lower Cretaceous Kootenay Formation, southern Rocky Mountains, Canada. *Bull. geol. Soc. Am.* **83**, 3199–3222.

Jefferies, R. P. S. and Minton, P. 1965. The mode of life of two Jurassic species of *Posidonia* (Bivalvia). *Palaeontol.* **8**, 156–185.

Jeletsky, J. A. 1963. *Malayomaorica* gen. nov. (Family Aviculopectinidae) from the Indo-Pacific Upper Jurassic; with comments on related forms. *Palaeontology* **6**, 148–160.

Jeletzky, J. A. 1965. Late Upper Jurassic and early Lower Cretaceous fossil zones of the Canadian Western Cordillera. *Bull. Geol. Surv. Canada* **103**.

Jeletsky, J. A. 1966. Upper Volgian (late Jurassic) ammonites and Buchias of Arctic Canada. *Bull. Geol. Surv. Canada* **128**.

Jenkyns, H. C. 1970 a. Growth and disintegration of a carbonate platform. *N. Jb. Geol. Paläont. Mh.* **6**, 325–344.

Jenkyns, H. C. 1970 b. Submarine volcanism and the Toarcian iron pisolites of Western Sicily. *Ecl. geol. Helv.* **63**, 549–572.

Jenkyns, H. C. 1970 c. Fossil manganese nodules from the west Sicilian Jurassic *Ecl. geol. Helv.* **63**, 741–774.

Jenkyns, H. C. 1971 a. Speculations on the genesis of crinoidal limestones in the Tethyan Jurassic. *Geol. Rundsch.* **60**, 471–488.

Jenkyns, H. C. 1971 b. The genesis of condensed sequences in the Tethyan Jurassic. *Lethaia* **4**, 327–352.

Jenkyns, H. C. 1972. Pelagic 'oolites' from the Tethyan Jurassic. *J. Geol.* **80**, 21–33.

Jenkyns, H. C. 1974. Origin of red nodular limestones (Ammonitico Rosso, Knollenkalk) in the Mediterranean Jurassic; a diagenetic model. In: K. J. Hsü and H. C. Jenkyns (editors), *Pelagic Sediments: On Land and Under the Sea.* Int. Ass. Sedimentol. Spec. Publ. 1, 249–271

Jenkyns, H. C. and Torrens, H. S. 1971. Palaeogeographic evolution of Jurassic seamounts in Western Sicily. *Ann. Inst. Geol. Publ. Hungar.* **54**, 91–104.

Jones, D. L., Bailey, E. H. and Imlay, R. W. 1969. Structural and stratigraphic significance of the *Buchia* zones in the Colyear Springs – Paskenta area, California. *U.S. Geol. Surv. Prof. Paper* 647-A.

Jordan, R. 1971 a. Zur Salinität des Meeres im höheren Oberen Jura Nordwest-Deutschlands. *Z. Deutsch. Geol. Ges.* **122**, 231–241.

Jordan, R. 1971 b. Megafossilien des Juras aus den Antalo-Kalk von Nord-Äthiopien. *Beih. geol. Jb.* **116**, 141–172.

Jordan, R. and Stahl, W. 1969. General considerations on isotopic palaeotemperature determinations and analyses on Jurassic ammonites. *Earth Planet. Sci. Lett.* **6**, 173–178.

Jordan, R. and Stahl, W. 1970. Isotopische Paläotemperatur Bestimmungen an jurassischen Ammoniten und grundsätzliche Voraussetzungen für diese Methode. *Geol. Jb.* **89**, 33–62.

Jung, W. 1963. Die Mesozoischen Sedimente am Sudostrand des Gotthard Massivs (zwischen Plaun la Greina und Versam). *Ecl. geol. Helv.* **56**, 655–754.

Jurgan, H. 1969. Sedimentologie des Lias der Berchtesgadener Kalkalpen. *Geol. Rundsch.* **58**, 464–501.

Jurkiewiczowa, I. 1967. The Lias of the western part of the Mesozoic zone surrounding the Swietokrzyskie (Holy Cross) Mountains and its correlation with the Lias of the Cracow–Wielun Range. *Biul. Inst. Geol.* **200**, 5–132.

Kaptarenko-Tschernousowa, O. K. 1964. Versuch eines stratigraphischen Vergleiches der Jura – Ablagerungen auf Grund ihrer Foraminiferen – Fauna. *Publ. Inst. Grand-Ducal Sect. Sci. Nat., Phys., Math. Luxembourg*, 429–437.

Karaszewski, W. 1962. The stratigraphy of the Lias in the northern Mesozoic zone surrounding the Swietz Krzyz Mountains (Central Poland). *Pr. Inst. Geol. Warsaw* **30**, cz. 3, 334–416. (In Polish, with English summary.)

Kauffman, E. G. 1967. Coloradan macroinvertebrate assemblages, central western interior, United States. *Sympos. Colorado School of Mines*, 67–143.

Kent, P. E. 1967. Outline geology of the southern North Sea Basin. *Proc. Yorks. geol. Soc.* **36**, 1–22.

Kent, P. E., Hunt, J. A. and Johnstone, D. W. 1971. The geology and geophysics of coastal Tanzania. *I.G.S. Geophys. Paper* No. 6.

Keulegan, G. H. and Krumbein, W. H. 1949. Stable configuration of bottom slope in a shallow sea and its bearing on geological processes. *Trans. Am. Geophys. Union* **30**, 855–861.

Khudoley, K. M. and Meyerhoff, A. A. 1971. Paleogeography and geological history of the Greater Antilles. *Geol. Soc. Am. Mem.* **129**, 1–99.

Klein, G. de Vries, 1965. Dynamic significance of primary structures in the Middle Jurassic Great Oolite Series, southern England. *Soc. Econ. Pal. Min. Spec. Publ.* **12**, 173–190.

Klinger, H. C., Kennedy, W. J. and Dingle, R. V. 1972. A Jurassic ammonite from South Africa. *N. Jb. Geol. Paläont. Mh.* **11**, 653–659.

References

Klohn Giehm, C. 1960. Geologia de la Cordillera de los Andes de Chile central, provincias de Santiago, O'Higgins, Colchagua y Curico. *Inst. Inves. Geol. Chile.* Ser. B, **8**.

Klüpfel, W. 1917. Über die sedimente der Flachsee im lothringer Jura. *Geol. Rundsch.* **7**, 97–109.

Kneuper-Haack, F. 1966. Ostracoden aus dem Wealden der Sierra de los Cameros (nordwestliche Iberische Ketten). *Beih. Geol. Jb.* **44**, 165–209.

Knox, R. W. O'B. 1970. Chamosite ooliths from the Winter Gill ironstone (Jurassic) of Yorkshire, England. *J. sedim. Petrol.* **40**, 1216–1225.

Knox, R. W. O'B. 1973. The Eller Beck Formation (Bajocian) of the Ravenscar Group of N.E. Yorkshire. *Geol. Mag.* **110**, 511–534.

Kobayashi, T. 1957. A trigonian faunule from Mindoro in the Philippine Islands. *J. Fac. Sci. Univ. Tokyo* **10**, 251–365.

Kobayashi, T. 1960. Notes on the geologic history of Thailand and adjacent territories. *Jap. J. Geol. Geog.* **31**, 129–148.

Komalarjun, P. and Sato, T. 1964. Aalenian (Jurassic) ammonites from Mae Sot, northwestern Thailand. *Jap. J. Geol. Geog.* **35**, 149–161.

Kopik, J. 1962. Faunistic criteria of stratigraphical subdivision of the Lias in north-western and central Poland. *Polsk. Akad. Nauk. Warsz.* 271–302. (In Polish, with English summary.)

Kopik, J., Bielecka, W., Styk, O. and Pazdrowa, O. 1967. The Middle and Upper Jurassic of the Czestochowa–Zawiercie sedimentary basin (the Cracow–Czestochowa Jura). *Biul. Inst. Geol.* **211**, 93–186.

Kotanski, Z. 1961. Tectogenèse et reconstitution de la paléogeographie de la zone haut-tatrique dans les Tatras. *Acta Geol. Pol.* **11**, 187–476. (Polish, with French summary.)

Krimholz, G. 1964. Sur la subdivision du Jurassique marine adoptée en U.R.S.S. *Publ. Inst. Grand Ducal Sect. Sci. Nat. Phys. Math. Luxenbourg,* 747–762.

Krimholz, G. 1972 (editor). *Jurassic System.* Vol. 10 of *Stratigraphy*, U.S.S.R. Minist. Geol. Moscow. (In Russian.)

Krömmelbein, K. 1956. Die ersten marine Fossilien (Trigoniidae, Lamellibr), aus der Cayetano-Formation West-Cubas. *Senck. leth.* **37**, 331–335.

Krömmelbein, K. 1962. Beiträge zur geologischen Kenntnis der Sierra de los Organos (Cuba). *Z. deutsch. geol. Ges.* **114**, 92–120.

Krömmelbein, K. and Wenger, R. 1966. Sur quelques analogies remarquables dans les microfaunes crétacées du Gabon et du Bresil oriental (Bahia et Sergipe). In: *Bassins sedimentaire du littoral africain, Part 1*, Union Int. Sci. Geol. Ass. Serv. Geol. Afr., Paris, 193–196.

Ksiazkiewicz, M. 1956. The Jurassic and Cretaceous of Bachowice (Western Carpathians). *Acta Geol. Pol.* **24**, 121–405. (In Polish, with English summary.)

Kuenan, P. H. 1950. *Marine geology.* John Wiley, New York.

Kutek, J. 1961. Le Kimeridgien et le Bononien de Stodnica. *Acta Geol. Pol.* **11**, 103–183.

Kutek, J. 1962. Le Kimeridgien supérieur et le Volgien inférieur de la bordure mesozoique nord-ouest des monts de Sainte Croix. *Acta Geol. Pol.* **12**, 445–527. (In Polish, with French summary.)

Kutek, J. 1967. Some problems of Upper Jurassic stratigraphy in Poland. *Biul. Inst. Geol.* **203**, 87–114.

Kutek, J. and Glazek, J. 1972. The Holy Cross area, central Poland, in the Alpine cycle. *Acta Geol. Pol.* **22**, 603–653.

Larsen, G. 1966. Rhaetic–Jurassic–Lower Cretaceous sediments in the Danish Embayment. *Danm. Geol. Undersøg.* Ser. 2 No. 91.

Larsen, R. L. and Chase, C. G. 1972. Late Mesozoic evolution of the western Pacific Ocean. *Bull. geol. Soc. Am.* **83**, 3627–3644.

Larsen, R. L. and Pitman, W. C. 1972. World-wide correlation of Mesozoic magnetic anomalies, and its implications. *Bull. geol. Soc. Am.* **83**, 3645–3662.

Laubscher, H. P. 1971. Das Alpen–Dinariden–Problem und die Palinspastik der sudlichen Tethys. *Geol. Rundsch.* **60**, 813–833.

Laughton, A. S., Sclater, J. G. and McKenzie, D. P. 1973. The structure and evolution of the Indian Ocean. In: D. H. Tarling and S. K. Runcorn (editors), *Implications of Continental Drift to the Earth Sciences* Vol. 1, Academic Press, New York, pp. 203–212.

Lefavrais-Raymond, A. and Horon, O. 1961. Bassin de Paris. *Colloque Lias français, B.R.G.M. Mém.* **4**, 3–56.

Lehner, P. 1969. Salt tectonics and Pleistocene stratigraphy on continental slope of northern Gulf of Mexico. *Bull. Am. Ass. Petrol. Geol.* **53**, 2431–2479.

Liechti, P., Roe, F. W., Haile, N. S. and Kirk, H. J. G. 1960. The geology of Sarawak, Brunei and the western part of North Borneo. *Bull. Geol. Surv. Dept. Brit. Terr. Borneo* **3**.

Lisitsyn, A. P. 1967. Basin relationships in distribution of modern siliceous sediments and their connection with climatic zonation. *Int. Geol. Rev.* **9**, 1114–1130.

Logan, B. W., Rezak, R. and Ginsburg, R. N. 1964. Classification and environmental significance of algal stromatolites. *J. Geol. 1964.* **72**, 68–83.

Longinelli, A. 1969. Oxygen-18 variations in belemnite guards. *Earth Planet. Sci. Lett.* **7**, 209–212.

Lotze, F. 1964. The distribution of evaporites in space and time. In: A. E. M. Nairn (editor), *Problems in Palaeoclimatology*, John Wiley, New York, pp. 491–507.

Luyendyk, B. P., Forsyth, D. and Phillips, J. D. 1972. Experimental approach to the paleocirculation of the oceanic surface waters. *Bull. geol. Soc. Am.* **83**, 2649–2666.

McDougall, I. 1963. Potassium–argon age measurements on dolerites from Antarctica and South Africa. *J. geophys. Res.* **68**, 1535.

McDougall, I. and Rüegg, N. R. 1966. Potassium–argon age measurements on the Serra Geral Formation of South America. *Geochim. Cosmochim. Acta.* **30**, 191.

McElhinny, M. W. 1970. Formation of the Indian Ocean. *Nature* **228**, 977–979.

McIver, N. L. 1972. Cenozoic and Mesozoic stratigraphy of the Nova Scotia Shelf. *Can. J. Earth Sci.* **9**, 54–71.

McKee, E. D. *et al.* 1956. *Palaeotectonic maps of the Jurassic system.* U.S. Geol. Surv. Misc. Geol. Inv. Map I, 175.

McKenzie, D. and Sclater, J. G. 1971. The evolution of the Indian Ocean since the late Cretaceous. *Geophys. J.* **24**, 437–528.

McKerrow, W. S., Johnson, R. T. and Jakobson, M. E. 1969. Palaeoecological studies in the Great Oolite at Kirtlington, Oxfordshire. *Palaeontology* **12**, 56–83.

References

McWhae, J. R. H., Playford, P. E., Lindner, A. W., Glenister, B. F. and Balne, B. E. 1958. The stratigraphy of Western Australia. *J. geol. Soc. Austral.* **4**, 1–161.

Mahadevan, C. and Srivamadas, A. 1958. The Gondwanas of the east coast of India. *20th Int. Geol. Cong. Mexico, Com. Corr. Sist. Karroo,* 105–112.

Malinowska, L. 1967. Biostratigraphy of Lower and Middle Oxfordian in the margin of the Swietokrzyskie Mountains. *Biul. Inst. Geol.* **209**, 53–112. (In Polish with English summary.)

Malinowska, L. 1972. Middle and Upper Oxfordian in the north-west part of the Czestochowa Jurassic. *Biul. Inst. Geol.* **233**, 6–67. (In Polish with English summary.)

Manton, W. I. 1968. The origin of associated basic and acid rocks in the Lebombo–Nuanetsi Igneous Province, southern Africa, as implied by strontium isotopes. *J. Petrol.* **9**, 23–39.

Martin, A. J. 1967. Bathonian sedimentation in southern England. *Proc. Geol. Ass. Lond.* **78**, 473–488.

Martinis, B. and Visintin, V. 1966. Données géologique sur le bassin sédimentaire côtier de Tafaya (Maroc meridional). In: D. Reyre (editor), *Bassins sedimentaire du littoral africain*, Ass. Serv. géol. africains, sympos., Paris, pp. 13–26.

Martinson, G. G. 1964. Significance of fresh water mollusca for the stratigraphy of Jurassic continental deposits of Asia. *Publ. Inst. Grand-Ducal, Sect. Sci. Nat., Phys. Math. Luxembourg,* 459–463.

Maubeuge, P. L. 1972. La carrière de Malancourt (Moselle): une contribution à la sédimentation et la stratigraphie du Bajocien moyen lorrain. *Bull. Ac. Soc. Lorraines des Sciences* **11**, No. 4.

Maync, W. 1966. Microbiostratigraphy of the Jurassic of Israel. *Bull. Israel Geol. Surv.* **40**.

Maxwell, A. E. *et al.* 1970. Deep sea drilling in the South Atlantic. *Science* **168**, 1047–1059.

Meischner, K. D. 1964. Allodapische Kalke, Turbidite in Riff-Nahen sedimentations-becken. In: A. H. Bouma and A. Brouwer (editors), *Turbidites*, Elsevier, Amsterdam, pp. 156–191.

Melville, R. V. 1956. The stratigraphical palaeontology, ammonites excluded, of the Stowell Park borehole. *Bull. geol. Surv. G.B.* No. 11, 67–139.

Menard, H. W. 1964. *Marine geology of the Pacific.* McGraw Hill, New York.

Meña Rojas, E. 1960. El Jurasico marino de la region de Cordoba. *Bol. Assoc. Mex. Geol. Petrol.* **12**, 243–252.

Mensink, H. 1967. Mariner Jura im westlichen Hindukusch (Afghanistan). *Geol. Rundsch.* **56**, 812–818.

Merla, G. 1952. Geologia dell' Appeninno settentrionale. *Bol. Soc. geol. Ital.* **70**, 95–382.

Minato, M., Gorai, M. and Hunahashi, M. 1965. *The geologic development of the Japanese Islands*, Tsukiji Shokan, Tokyo, p. 442.

Misik, M. 1964. Lithofazielles studium des Lias der Grossen Fatra und des westlichen Teils der Niederen Tatra. *Sbornik Geol. Vied. Zapadne Karpaty.* Rad 2K, zv. 1, 7–92.

Misik, M. 1966. *Microfacies of the Mesozoic and Tertiary limestones of the west Carpathians.* Sloven. Akad. Vied., Bratislava.

Misik, M. and Rakus, M. 1964. Bemerkungen zu raumlichen Beziehungen des Lias und zur Paläogeographie des Mesozoikum in der Grossen Fatra. *Sbornik Geol. Vied. Zapadne Karpaty*, Rad 2H, zv. 1, 159–199.

Moberley, R. M. J. 1960. Morrison, Cloveley and Sykes Mountain formations, northern Bighorn Basin, Wyoming and Montana. *Bull. geol. Soc. Am.* **71**, 1137–1176.

Montanari, L. and Crespi, U. 1974. Eventi Domeriano–Toarciani nelle Prealpi Lombarde Occidentale. *Att. Ist. Geol. Univ. Pavia* **24**, 92–119.

Monty, C. 1965. Recent algal stromatolites in the windward lagoon, Andros Island, Bahamas. *Ann. Soc. géol. Belg. Bull.* **88**, 269–276.

Moore, D. G. 1970. Reflection profiling studies of the California continental borderland: structure and Quaternary turbidite basins. *Geol. Soc. Am. Spec. Paper* 107.

Morton, D. M. 1959. Geology of Oman. *Proc. 5th World Petrol. Congr.* Sect. 1, 277–294.

Morton, N. 1965. The Bearreraig Sandstone Series (Middle Jurassic) of Skye and Raasay. *Scot. J. Geol.* **1**, 189–216.

Mouterde, R. 1967. Le Lias du Portugal: vue d'ensemble et division en zones. *Comm. Serv. geol. Portugal* **52**, 209–226.

Mouterde, R., *et al.* 1971 a. Les zones du Jurassique en France. *C.R. Somm. Séances Soc. géol. France*, fasc. 6, 1–27.

Mouterde, R., Ramalho, M., Rocha, R. B., Ruget, C. and Tintant, H. 1971 b. Le Jurassique du Portugal: esquisse stratigraphique et zonale. *Bol. Soc. geol. Portugal.* **18**, 73–104.

Müller, G. and Blaschke, R. 1969. Zur Entstehung des Posidonienschiefers (Lias ε) *Naturwiss.* **12**, 635–636.

Mutihac, V. 1971. Les facies du Jurassique de la zone centrale des Carpates orientaux (Roumanie). *Ann. Inst. Geol. Publ. Hungar.* **54**, 185–194.

Nachev, I. 1966. The Jurassic System in the north-eastern part of the Kraishte. *Rev. Bulgar. geol. Soc.* **27**, 85–90. (In Russian, with English summary.)

Nachev, I., Sapunov, I. G. and Stephanov, J. 1963. Stratigraphy and lithology of the Jurassic system between Gorno Ozirovo and Prevala villages (North-West Bulgaria). *Bulgar. Akad. Nauk. Geol. Inst. Tr. Geol. Bulgar. Ser. Pol.* **5**, 99–146.

Nalivkin, D. V. 1960. *The geology of the U.S.S.R.* Pergamon, Oxford.

Neale, J. W. 1967. Ostracods from the type Berriasian (Cretaceous) of Berrias (Ardèche, France) and their significance. *Univ. Kansas Dept. Geol. Spec. Publ.* 2, 539–569.

Neaverson, E. 1925. The petrography of the Upper Kimmeridge Clay and Portland Sand in Dorset, Wiltshire, Oxfordshire and Buckinghamshire. *Proc. Geol. Ass. Lond.* **36**, 240–256.

Neumayr, M. 1883. Über klimatische Zonen während der Jura – und Kreidzeit. *K.Akad. Wiss. Wien Denkschr. Math.-naturh. Kl.* **47**, 277–310.

Newell, N. D. 1967. Revolutions in the history of life. *Geol. Soc. Am. Spec. Paper* 89, 63–91.

Noël, D. 1965. *Sur les Coccolithes du Jurassique Européan et d'Afrique du Nord.* Edit. Centre Nat. Rech. Scient. Paris.

Noël, D. 1972. Nannofossiles calcaires de sediments jurasiques finement lamines. *Bull. Mus. Nat. d'Hist. Nat.* 3/75, 95–156.

References

Norling, E. 1972. Jurassic stratigraphy and foraminifera of western Scania, Southern Sweden. *Sver. Geol. Undersök, Avh. Uppsat. Ser. Ca*, No. 47.

Norris, G. 1965. Triassic and Jurassic miospores and acritarchs from the Beacon and Ferrar Groups, Victoria Land, Antarctica. *N.Z. J. Geol. Geophys.* **8**, 236–277.

Norris, G. 1969. Miospores from the Purbeck Beds and marine Upper Jurassic of southern England. *Palaeontology* **12**, 574–620.

Nørvang, A. 1957. The Foraminifera of the Lias Series in Jutland, Denmark. *Medd. Dansk. Geol. Foren.* **13**, 279–413.

Oertel, G. 1956. Transgressionen im Malm der portugiesischen Estremadura. *Geol. Rundsch.* **45**, 304–313.

Ørvig, T. 1960. The Jurassic and Cretaceous of Andøya in northern Norway. *Norges Geol. Undersøk.* **208**, 344–350.

Palmer, A. R. 1973. Cambrian trilobites. In: A. Hallam (editor), *Atlas of Palaeobiogeography*, Elsevier, Amsterdam, pp. 3–11.

Palmer, T. J. and Jenkyns, H. C. 1975. A carbonate island barrier from the Great Oolite of central England. *Sedimentology* **22**, 125–135.

Palmer, T. J. and Fürsich, F. T. 1974. The ecology of a Middle Jurassic hardground and crevice fauna. *Palaeontology* **17**, 507–524.

Pannekoek, A. J. 1956 (editor). *Geological history of the Netherlands*. Govt. Printing Office, The Hague.

Paquet, J. 1969. Etudes géologiques de l'ouest de la province de Murcie (Espagne). *Mém. Soc. géol. France* No. 111, N.S. **48**.

Parker, J. R. 1967. The Jurassic and Cretaceous sequence in Spitsbergen. *Geol. Mag.* **104**, 487–505.

Parsons, C. F. 1974. On the Sowerbyi and so-called Sowerbyi zones of the Middle Bajocian. *News Letters in Stratigraphy* **3**, 153–180.

Passerini, P. 1964. Il Monte Cetona (Provincia di Sierra). *Boll. Soc. geol. Ital.* **83**, 223–338.

Passendorfer, E. 1961. Evolution paléogéographique des Tatras. *Acta Geol. Pol.* **30**, 351–387. (In Polish, with French summary.)

Patrulius, D. and Popa, E. 1971. Lower and Middle Jurassic ammonite zones in the Rumanian Carpathians. *Ann. Inst. Geol. Publ. Hungar.* **54**, 131–148.

Pavia, G. 1971. Ammoniti del Baiociano superiore di Digne (Francia SE, Dip. Basses Alpes). *Boll. Soc. paleont. Ital.* **10**, 75–142.

Pektović, K., Marković, B. Veselinović, D., Andjelković, D. and Pasić, M. 1960. Das Mesozoikum Jugoslawiens. *Ann. Inst. Publ. Hungar.* **49**, 201–261.

Peña Muñoz, M. J. 1964. Amonitas del Jurasico superior y del Cretacico Inferior del extremo oriental del Estado de Durango, Mexico. *Paleont. Mexicana* No. 20.

Perez-Ibarguengoitia, J. M., Hokuto-Castilla, A. and Cserna, Z. 1965. Estratigrafia y paleontologia del Jurasico superior de la parte contro meridional del Estado de Puebla. *Paleont. Mexicana*. No. 21.

Peterson, J. A. 1957. Marine Jurassic of northern Rocky Mountains and Williston Basin. *Bull. Am. Ass. Petrol. Geol.* **41**, 399–440.

Peterson, J. A. 1972. Jurassic System. In: *Geological atlas of the Rocky Mountains*, Rocky Mtn. Assoc. Geologists, pp. 177–189.

Pinna, G. 1966. Ammoniti del Lias superiore (Toarcian) dell' Alpe Turati (Erba, Como). *Mem. Soc. Ital. Sci. nat. mus. cir. St. nat. Milano* **14**, 85–136.

Pipiringos, G. N. 1968. Correlation and nomenclature of some Triassic and Jurassic rocks in south-central Wyoming. *U.S. Geol. Surv. Prof. Paper* 594-D.

Pitman, W. C. and Talwani, M. 1972. Sea floor spreading in the North Atlantic. *Bull. geol. Soc. Am.* 83, 619–46.

Playford, P. E. 1959. Jurassic stratigraphy of the Geraldton district, Western Australia. *J. Roy. Soc. W. Austral.* 42, 101–124.

Playford, G. and Dettmann, M. E. 1965. Rhaeto-Liassic plant mirofossils from the Leigh Creek Coral Measures, South Australia. *Senck. leth.* 46, 127–181.

Plumstead, E. P. 1964. Palaeobotany of Antarctica. In: R. J. Adie (editor), *Antarctic Geology*, North-Holland, Amsterdam, 637–654.

Poole, E. G. and Whiteman, A. J. 1966. *Geology of the country around Nantwich and Whitchurch.* Mem. geol. Surv. G.B.

Poole, E. G., Williams, B. J. and Hains, B. A. 1968. *Geology of the country around Market Harborough.* Mem. geol. Surv. G.B.

Poole, F. G. 1964. Palaeowinds in the Western United States. In: A. E. M. Nairn (editor), *Problems in Palaeoclimatology*, Interscience, London and New York, pp. 394–405.

Porrenga, D. H. 1965. Chamosite in Recent sediments of the Niger and Orinoco deltas. *Geol. Mijnb.* 44, 400–403.

Powers, R. W. 1962. Arabian Upper Jurassic carbonate reservoir rocks. *Mem. Am. Ass. Petrol. Geol.* 1, 122–192.

Powers, R. W., Ramirez, L. F., Redmond, C. D. and Elberg, E. L. 1966. Geology of the Arabian Peninsula: sedimentary geology of Saudi Arabia. *U.S. Geol. Surv. Prof. Paper* 560-D.

Praturlon, A. 1966. Algal assemblages from Lias to Paleocene in southern Latium–Abruzzi: a review. *Boll. Soc. geol. Ital.* 85, 167–194.

Pugh, M. E. 1968. Algae from the Lower Purbeck Limestones of Dorset. *Proc. Geol. Ass. Lond.* 79, 513–523.

Pulfrey, W. 1963. Kenya. *Lex. Strat. Internat.*, Vol. 4 (editors D. T. Donovan and J. E. Hemingway), fasc. 8a.

Purser, B. H. 1969. Syn-sedimentary marine lithification of Middle Jurassic limestones in the Paris Basin. *Sedimentology* 12, 205–230.

Querol, R. 1969. Petroleum exploration in Spain. In: P. Hepple (editor), *The exploration for petroleum in Europe and North Africa.* Inst. Petrol. London.

Quilty, P. G. 1970. Jurassic ammonites from Ellsworth Land, Antarctica. *J. Paleont.* 44, 110–116.

Raab, M. 1962. Jurassic–early Cretaceous ammonites from the southern coastal plain of Israel. *Bull. geol. Surv. Israel* 34, 24–30.

Radoičić, R. 1966. Microfacies du Jurassique des Dinarides externes de la Yougoslavie. *Geol. Razp. Poročila* 9.

Radwanski, A. and Szwlczewski, M. 1966. Jurassic stromatolites of the Villany Mountains (southern Hungary). *Ann. Univ. Scient. Budapest Eötvös, Nom. Sect. Geol.* 9, 87–107.

Raileanu, G., Patrulius, D., Bleahu, M., Nastaseanu, S. and Semaka, A. 1964. Observations sur les limites des séries jurassiques dans les Carpates Roumaines. *Publ. Inst. Grand-Ducal Sect. Sci. Nat., Phys. Math. Luxembourg*, 675–690.

Rakus, M. 1964. Paläontologische studien im Lias der Grossen Fatra und des

References

westlichen Teils der Niederen Tatra. *Sbornik. Geol. Vied. Zapadne Karpaty*, Rad 2K, zv. 1, 93–155.

Reineck, H. E. and Singh, I. B. 1971. Genesis of laminated sand and graded rhythmites in storm-sand layers of shelf mud. *Sedimentol.* 18, 123–128.

Remane, J. 1964. Untersuchungen zur systematik und stratigraphie der Calpionellen in der Jura-Kreide-Schichten des Vocontischen Troges. *Palaeontogr.* A 123, 1–57.

Renz, C. 1955. *Die vorneogene stratigraphie der normal sedimentaren Formationen Griechenlands.* Inst. geol. sub surface Res., Athens.

Reyment, R. A. 1959. On Liassic ammonites from Skane, southern Sweden. Stockholm Contr. Geol. 2, 103–156.

Reyre, D. 1966 (editor). *Bassins sedimentaire du littoral africain.* Assoc. Serv. geol. Africains. Symposium, Paris.

Rhoads, D. C. and Morse, J. W. 1971. Evolutionary and ecologic significance of oxygen-deficient marine basins. *Lethaia* 4, 413–428.

Rhoads, D. C. and Young, D. K. 1970. The influence of deposit-feeding organisms on sediment stability and community trophic structure. *J. mar. Res.* 28, 150–178.

Ridd, M. F. 1971. South-east Asia as a part of Gondwanaland. *Nature* 234, 531–533.

Robinson, P. L. 1971. A problem of faunal replacement on Permo-Triassic continents. *Palaeontology* 14, 131–153.

Roddick, J. A., Wheeler, J. O., Gabrielse, H. and Souther, J. G. 1966. Age and nature of the Canadian part of the circum-Pacific orogenic belt. *Tectonophys.* 4, 319–337.

Rohrlich, V., Calvert, S. E. and Price, N. B. 1969. Chamosite in the recent sediments of Lock Etive, Scotland. *J. sedim. Petrol.* 39, 624.

Romariz, C. 1960. Estudo geologico e petrografico da area tifonica do Soure. *Com. Serv. Geol. Portugal* 44, 1–219.

Romer, A. S. 1961. Palaeozoological evidence of climate. 1, Vertebrates. In: A. E. M. Nairn (editor), *Descriptive palaeoclimatology*, Interscience, London and New York.

Ronov, A. B. 1968. Probable changes in the composition of sea water during the course of geological time. *Sedimentol.* 10, 25–43.

Ronov, A. B. and Khain, V. Y. 1962. Jurassic formations of the world. *Sovetsk. Geol.* No. 1, 9–34 (see *Int. Geol. Review* 5, 812).

Rossi Ronchetti, C. and Fantini Sestini, N. 1961. La fauna giurassica di Karkar (Afghanistan). *Riv. Ital. Paleont. Strat.* 67, 103–140.

Rothpletz, A. 1910. Über die Einbettung der Ammoniten in den Solnhofener Schichten. *Abh. Bayer. Akad. Wiss. 2 Kl.* 24.

Rouselle, L. 1965. Rhynchonellidae, Terebratulidae et Zeilleridae du Dogger marocain. *Trav. Inst. Scient. Chérif. Sér. Géol. Géog. Phys.* No. 13.

Rüegg, W. 1957. Geologie zwischen Cañete–San Juan 13° 00′ – 15° 24′ sud Peru. *Geol. Rundsch.* 45, 775–858.

Ruget-Perrot, C. 1961. Etudes stratigraphiques sur le Dogger et le Malm inférieur du Portugal au Nord du Tage. *Serv. geol. Portugal Mém.* No. 7.

Ruiz, C., Aguirre, E., Corvalan, J., Rose, H. J., Segerstrom, K. and Stern, T. W. 1961. Ages of batholithic intrusions of northern and central Chile. *Bull. geol. Soc. Am.* 72, 1551–1560.

Rusbult, J. and Petzka, M. 1964. Zur stratigraphie des Lias in NE-Mecklenburg. *Ber. geol. Ges. DDR* **9**, 625–634.

Ryther, J. H. 1963. Geographic variations in productivity. In: M. N. Hill (editor), *The Sea*, Vol. 2, Interscience, London & New York, pp. 347–380.

Saggerson, E. P. and Miller, J. M. 1957. Geology of the Tekabba–Wergudud area, Mandera District. *Rep. Geol. Surv. Kenya* **40**.

Saks, V. N. 1972 (editor). *Jurassic–Cretaceous boundary and Berriasian stage in boreal realm*. Acad. Sci. U.S.S.R. Inst., Geol. Novosibirsk.

Saks, V. N. and Strelkov, S. A. 1961. Mesozoic and Cenozoic of the Soviet Arctic. In: G. O. Raasch (editor), *Geology of the Arctic*, Vol. 1, Toronto, pp. 48–67.

Saks, V. N., Mesezhnikov, M. S. and Shulgina, N. I. 1964. About the connection of the Jurassic and Cretaceous marine basins in the north and south of Eurasia. *Dok. Sovetsk. Geol. Mezh. Geol. Kong. 22nd Sess.* 163–174. (In Russian.)

Sanders, H. L. 1968. Marine benthic diversity: a comparative study. *Amer. Nat.* **102**, 243–282.

Sapunov, I. 1959. Stratigraphical and palaeontological studies of the Toarcian in the vicinity of the town of Tetevan (Central Balkan Range). *Bull. geol. Inst., Bulg. Ac. Sci., Ser. Pal.* **1**, 17–41. (In Russian, with English summary.)

Sapunov, I. G. 1969. On certain recent stratigraphical problems of the Jurassic system in Bulgaria. *Bull. geol. Inst., Bulg. Ac. Sci., Ser. Strat.* **18**, 5–20. (In Russian, with English summary.)

Sapunov, I. G. and Stephanov, J. 1964. The stages, substages, ammonite zones and subzones of the lower and Middle Jurassic in the Western and Central Balkan Range (Bulgaria). *Publ. Inst. Grand-Ducal Sect. Sci. Nat., Phys. Math. Luxembourg*, 705–718.

Sapunov, I. G., Tchoumatchenco, P. V. and Shopov, V. L. 1967. Biostratigraphy of the Lower Jurassic rocks near the village of Komshtitsa, district of Sofia (Western Balkan Range). *Bull. geol. Inst., Bulg. Ac. Sci., Ser. Strat.* **16**, 125–143. (In Russian, with English summary.)

Sarjeant, W. A. S. 1964. The stratigraphic application of fossil microplankton (Dinoflagellates and Hystrichospheres) in the Jurassic. *Publ. Inst. Grand-Ducal Sect. Sci. Nat., Phys. Math. Luxembourg*, 441–448.

Sato, T. 1960. A propos des courants océaniques froids prouves par l'existence des ammonites d'origine arctique dans le Jurassique Japonais. *21st Int. geol. Congr.* Part 12, 165–169.

Sato, T. 1962. Etudes biostratigraphiques des ammonites du Jurassique du Japon. *Mém. Soc. géol. France* **94**.

Sato, T. 1964. Le Jurassique du Japon – zones d'ammonites. *Publ. Inst. Grand-Ducal. Sect. Sci. Nat., Phys. Math. Luxembourg*, 885–896.

Sato, T., Hayami, I., Tamura, M. and Maeda, S. 1963. The Jurassic. In: F. Takai, T. Matsumoto and R. Toriyama (editors), *Geology of Japan*, Tokyo.

Schaeffer, B. 1971. Mesozoic fishes and climate. *Proc. N. Am. Paleont. Conv. Sept. 1969, Chicago*, Part D, 376–388.

Scheibnerova, V. 1972. Foraminifera and their Mesozoic biogeoprovinces. *24th Int. Geol. Congr.* Sect. 7, 331–338.

Schellmann, W. 1969. Die Bildungsbedingungen sedimentärer Chamosit – und Hämatit-Eisenerze am Beispiel der Lagerstätte Echte. *N. Jb. Miner. Abh.* **111**, 1–31.

References

Schindewolf, O. H. 1957. Über den Lias von Peru. *Geol. Jb.* **74**, 151–159.

Schirardin, J. 1955. Contribution à la stratigraphie et paléontologie de l'Oxfordien moyen et supérieur de la basse Alsace. *Bull. Serv. Carte Geol. Alsace–Lorraine* B**8**, 21–59.

Schlager, W. and Schlager, M. 1973. Clastic sediments associated with radio larites (Tanglboden-Schichten, Upper Jurassic, Eastern Alps). *Sedimentol.* **20**, 65–90.

Schloz, W. 1968. Über Beobachtungen zur Ichnofazies und über umgelagerte Rhizocorallien im Lias Schwabens. *N. Jb. Geol. Paläont. Mh.* **11**, 691–698.

Schloz, W. 1972. Zur Bildungsgeschichte der Oolithenbank (Hettangium) in Baden–Württemberg. *Arb. Inst. Geol. Paläont. Univ. Stuttgart*, N.S. **67**, 101–212.

Schmalz, R. F. 1969. Deep-water evaporite deposition: a genetic model. *Bull. Am. Ass. Petrol. Geol.* **53**, 798–823.

Schumacher, K. H. and Sonntag, H. 1964. Zur Stratigraphie und Ausbildung des Lias im Norden der Deutschen Demokratischen Republik. *Geologie* **13**, 303–315.

Schwarzbach, M. 1964. *Climates of the Past*. Enke, Stuttgart.

Segerstråle, S. G. 1957. Baltic Sea. *Geol. Soc. Am. Mem.* 67, Vol. 1, 751–802.

Seilacher, A. 1963. Umlagerung und Rolltransport von Cephalopoden–Gehäusen. *N. Jb. Geol. Paläont. Mh.* **11**, 593–615.

Seilacher, A. 1967. Bathymetry of trace fossils. *Mar. Geol.* **5**, 413–428.

Seilacher, A., Drozdzewski, G. and Haude, R. 1968. Form and function of the stem in a pseudoplanktonic crinoid (*Seirocrinus*). *Palaeontology* **11**, 275–282.

Sellwood, B. W. 1970. The relation of trace fossils to small sedimentary cycles in the British Lias. In: T. P. Crimes and J. C. Harper (editors), *Trace fossils*, Seel House Press, Liverpool, pp. 489–584.

Sellwood, B. W. 1971. The genesis of some sideritic beds in the Yorkshire Lias (England). *J. sedim. Petrol.* **41**, 854–858.

Sellwood, B. W. 1972a. Tidal-flat sedimentation in the Lower Jurassic of Bornholm, Denmark. *Palaeogeog., Palaeoclimatol., Palaeoecol.* **11**, 93–106.

Sellwood, B. W. 1972b. Regional environmental changes across a Lower Jurassic stage-boundary in Britain. *Palaeontology* **15**, 125–157.

Sellwood, B. W. and Hallam, A. 1974. Bathonian volcanicity and North Sea rifting. *Nature* **252**, 27–28.

Sellwood, B. W. and Jenkyns, H. C. 1975. Basins and swells and the evolution of an epeiric sea (Pliensbachian–Bajocian of Great Britain). *J. geol. Soc. Lond.* **131**.

Sellwood, B. W. and McKerrow, W. S. 1974. Depositional environments in the lower part of the Great Oolite Group of Oxfordshire and north Gloucestershire. *Proc. Geol. Ass. Lond.* **85**, 189–210.

Sellwood, B. W., Durkin, M. K. and Kennedy, W. J. 1970. Field meeting on the Jurassic and Cretaceous rocks of Wessex. *Proc. Geol. Ass. Lond.* **81**, 715–732.

Semaka, A. 1961. Über die pflanzenführenden Liasschichten Rumäniens (2, Danubikum). *N. Jb. Geol. Paläont. Mh.* **8**, 389–394.

Seyed Emami, K. 1971. The Jurassic Badamu Formation in the Kerman region, with remarks on the Jurassic stratigraphy of Iran. *Geol. Surv. Iran* Rep. No. 19, 5–79.

Shaw, A. B. 1964. *Time in stratigraphy*. McGraw-Hill, New York.

Shearman, D. J. 1966. Origin of marine evaporites by diagenesis. *Trans. Inst. Min. Metall.* Sect. B, **75**, 207–216.

Shinn, E. A. 1968. Practical significance of birdseye structures in carbonate rocks. *J. sedim. Petrol.* **38**, 215–223.

Sirna, G. 1962. Stratigrafia e microfacies dei lembi mesozoici della valle di Galati Mamestino (Sicilia nord-orientale). *Geol. Rom.* **1**, 191–203.

Skwarko, S. K. 1970. Bibliography of the Mesozoic palaeontology of Australia and eastern New Guinea. *Bur. Min. Res. Canberra Bull.* **108**.

Smith, A. G. 1971. Alpine deformation and the oceanic areas of the Tethys, Mediterranean, and Atlantic. *Bull. geol. Soc. Am.* **82**, 2039–2070.

Smith, A. G. and Hallam, A. 1970. The fit of the southern continents. *Nature* **225**, 139–144.

Smith, A. G., Briden, J. C. and Drewry, G. E. 1973. Phanerozoic world maps. *Spec. Papers in Palaeont.* No. 12, 1–39.

Smithson, F. 1942. The Middle Jurassic rocks of Yorkshire: a petrological and palaeogeographical study. *Quart. J. Geol. Soc. Lond.* **98**, 27–59.

Société Chérif. de Petroles 1966. Le bassin du sud-ouest marocain. *Assoc. Serv. Géol. Africains, Sympos.*, Paris, pp. 5–12.

Sorgenfrei, Th. and Buch, A. 1964. Deep tests in Denmark 1935–1959. *Danm. Geol. Unders.* Ser. 3, No. 36.

Spaeth, C., Hoefs, J. and Vetter, U. 1971. Some aspects of isotopic composition of belemnites and related paleotemperatures. *Bull. geol. Soc. Am.* **82**, 3139–3150.

Spengler, A. de, Castelain, J., Cauvin, J. and Leroy, M. 1966. Le bassin secondaire–tertiaire du Senegal. *Assoc Serv. Géol. Africains, Sympos.*, Paris.

Srivastava, S. K. 1963. Polospores from Jurassics of Rayasthan, India. *Nature* **198**, 1323–1324.

Stanley, K. O. 1971. Tectonic and sedimentologic history of Lower Jurassic Sunrise and Dunlap formations, west-central Nevada. *Bull. Am. Ass. Petrol. Geol.* **55**, 454–477.

Stanley, K. O., Jordan, W. M. and Dott, R. H. 1971. New hypothesis of early Jurassic paleogeography and sediment dispersal for western United States. *Bull. Am. Ass. Petrol. Geol.* **55**, 10–19.

Stauffer, P. H. and Gobbett, D. J. 1972. Southeast Asia a part of Gondwanaland? *Nature Phys. Sci.* **240**, 139.

Stehli, F. G., McAlester, A. L. and Helsey, C. E. 1967. Taxonomic diversity of Recent bivalves and some implications for geology. *Bull. geol. Soc. Am.* **78**, 455–466.

Stevens, C. H. 1971. Distribution and diversity of Pennsylvanian marine faunas relative to water depth and distance from shore. *Lethaia* **4**, 403–412.

Stevens, G. R. 1965. The Jurassic and Cretaceous belemnites of New Zealand and a review of the Jurassic and Cretaceous belemnites of the Indo-Pacific region. *N.Z. geol. Surv. Paleont. Bull.* **36**.

Stevens, G. R. 1967. Upper Jurassic fossils from Ellsworth Land, West Antarctica, and notes on Upper Jurassic biogeography of the South Pacific region. *N.Z. J. Geol. Geophys.* **10**, 345–393.

Stevens, G. R. 1968. The Jurassic system in New Zealand. *N.Z. Geol. Surv. Rep.* 35.

References

Stevens, G. R. 1971. Relationship of isotopic temperatures and faunal realms to Jurassic–Cretaceous palaeogeography, particularly of the S.W. Pacific. *J. roy. Soc. N.Z.* 1, 145–158.

Stevens, G. R. 1973. Jurassic belemnites. In: A. Hallam (editor), *Atlas of Palaeobiogeography*, Elsevier, Amsterdam, pp. 259–274.

Stevens, G. R. and Clayton, R. N. 1971. Oxygen isotope studies on Jurassic and Cretaceous belemnites from New Zealand and their biogeographic significance. *N.Z.J. Geol. Geophys.* 14, 829–897.

Stipanicic, P. N. 1966. El Jurassico en Vega de la Veranada (Neuquen), el Oxfordense y el diastrofismo divesiano (Agassiz–Yaila) en Argentina. *Rev. Assoc. geol. Argentina* 20, 403–478.

Stipanicic, P. N. 1969. El avance en los conocimientos del Jurassico Argentino a partir del esquema de Groeber. *Rev. Assoc. geol. Argentina* 24, 367–388.

Stipanicic, P. N. and Reig, O. A. 1956. El Complejo porfirico de la Patagonia extraandina y su fauna de anuros. *Acta Geol. Lilloana* 1, 185–230.

Stipanicic, P. N. and Rodrigo, F. 1970. El diastrofismo Jurassico en Argentina y Chile. *Cuart. Jorn. geol. Argentinas* 2, 353–368.

Stöcklin, J. 1968. Structural history and tectonics of Iran: a review. *Bull. Am. Ass. Petrol. Geol.* 52, 1229–58.

Stöcklin, J., Eftekhar-Nezhad, J. and Hushmand-Zadeh, A. 1965. Geology of the Shotori Range (Tabas area, East Iran). *Geol. Surv. Iran Rep.* 3.

Strakhov, N. M. 1967. *Principles of Lithogenesis.* Vol. 1. (English translation of Russian text edited by S. I. Tomkeieff and J. C. Hemingway.) Oliver & Boyd, Edinburgh.

Sturani, C. 1962. Il complesso sedimentario autoctono all estremo nord-occidentale del massiccio dell Argentera (Alpi Merittime). *Mem. 1st Geol. Min. Univ. Padova* 22.

Sturani, C. 1964a. La successione delle faune ad ammoniti nelle formazioni mediogiurasiche della Prealpi Venete occidentale. *Mem. Ist. Geol. Min. Univ. Padova* 24.

Sturani, C. 1964b. Ammoniti mediogiurassiche del Veneto: Faune del Baiociano terminale (zone a Garantiana e a Parkinsoni). *Mem. Ist. Geol. Min. Univ. Padova* 24.

Sturani, C. 1966. Ammonites and stratigraphy of the Bathonian in the Digne–Barrème area (south-eastern France, Dept. Basses-Alpes). *Boll. Soc. Paleont. Ital.* 5, 3–57.

Sturani, C. 1971. Ammonites and stratigraphy of the 'Posidonia alpina' beds of the Venetian Alps. *Mem. 1st Geol. Min. Univ. Padova* 28, 190 p.

Surlyk, F. and Birkelund, F. 1972. The geology of southern Jameson Land. *Rapp. Grønl. geol. Unders.* 48, 61–74.

Surlyk, F., Callomon, J. H., Bromley, R. G. and Birkelund, T. 1973. Stratigraphy of the Jurassic–Lower Cretaceous sediments of Jameson Land and Scoresby Land, East Greenland. *Grønl. geol. Undersøg. Bull.* No. 105.

Sykes, R. M. 1975. The stratigraphy of the Callovian and Oxfordian in northern Scotland. *Scot. J. Geol.* 11, 1–28.

Sylvester Bradley, P. C. and Ford, T. D. 1968 (editors). *The geology of the East Midlands.* Leicester Univ. Press.

Szulczewski, M. 1965. Observation sur la genèse des calcaires noduleux des Tatras. *Ann. Soc. Geol. Pol.* **35**, 243–261. (In Polish, with French Summary.)

Szulczewski, M. 1968. Jurassic stromatolites of Poland. *Acta geol. Pol.* **18**, 1–99.

Takin, M. 1971. Continental drift in the Middle East. *Nature* **235**, 147–149.

Talbot, M. R. 1971. Calcite cements in the Corallian Beds (Upper Oxfordian) of southern England. *J. sedim. Petrol.* **41**, 261–273.

Talbot, M. R. 1972. The preservation of scleractinian corals by calcite in the Corallian beds (Oxfordian) of southern England. *Geol. Rundsch.* **61**, 731–741.

Talbot, M. R. 1973. Major sedimentary cycles in the Corallian Beds. *Palaeogeog., Palaeoclimatol., Palaeoecol.* **14**, 293–317.

Talbot, M. R. 1974. Ironstones in the Upper Oxfordian of Southern England. *Sedimentol.* **21**, 433–450.

Tan, F. C. and Hudson, J. D. 1970. Isotopic studies of the palaeoecology and diagenesis of the Great Estuarine Series (Jurassic) of Scotland. *Scot. J. Geol.* **10**, 91–128.

Tan, F. C., Hudson, J. D. and Keith, M. L. 1970. Jurassic (Callovian) paleotemperatures from Scotland. *Earth Planet. Sci. Lett.* **9**, 421–426.

Tanner, W. F. 1965. Upper Jurassic paleogeography of the Four Corners region. *J. sedim. Petrol.* **35**, 564–574.

Tarling, D. H. 1972. Another Gondwanaland. *Nature* **238**, 92–93.

Taylor, J. H. 1949. *Petrology of the Northampton Sand Ironstone Formation.* Mem. geol. Surv. G.B.

Taylor, J. H. 1963. *Geology of the country around Kettering, Corby and Oundle.* Mem. geol. Surv. G.B.

Teofilak-Maliszewska, A. 1967. The petrography of the Liassic deposits in the Polish lowland area. *Biul. Inst. Geol.* **207**, 67–155. (In Polish, with English summary.)

Termier, H. and Termier, G. 1960. *Atlas de Paléogéographie.* Masson, Paris.

Terris, A. P. and Bullerwell, W. 1965. Investigations into the underground structure of southern England. *Adv. Sci. Lond.* **22**, 232–252.

Thomas, W. A. and Mann, J. C. 1966. Late Jurassic depositional environments, Louisiana and Arkansas. *Bull. Am. Ass. Petrol. Geol.* **50**, 178–182.

Thompson, A. O. and Dodson, R. G. 1960. Geology of the Bur Mayo-Tarbaj area. *Rep. Geol. Surv. Kenya* **47**.

Tollman, A. 1963. *Ostalpensynthese.* Deuticke, Vienna.

Torrens, H. S. 1965. Revised zonal scheme for the Bathonian stage of Europe. *Rep. Carpatho-Balkan Geol. Ass. 7th Congr. Sofia.* Part 2, Vol. 1, pp. 47–55.

Torrens, H. S. 1969. Field meeting in the Sherborne–Yeovil district. *Proc. Geol. Ass. Lond.* **80**, 301–330.

Torrens, H. S. and Calloman, J. H. 1968. The Corallian Beds, the Ampthill Clay and the Kimmeridge Clay. In: P. C. Sylvester Bradley and T. D. Ford (editors), *The Geology of the East Midlands,* Leicester Univ. Press, pp. 291–299.

Tourtelot, H. A. and Rye, R. O. 1969. Distribution of oxygen and carbon isotopes in fossils of Late Cretaceous age, Western interior region of North America. *Bull. geol. Soc. Am.* **80**, 1903–1922.

Townson, W. G. 1971. Facies analysis of the Portland Beds. Unpubl. D.Phil. thesis Univ. Oxford.

References

Trümpy, R. 1960. Paleotectonic evolution of the Central and Western Alps. *Bull. geol. Soc. Am.* **71**, 843–908.

Trümpy, R. 1971. Sur le Jurassique de la zone Helvetique en Suisse. *Ann. Inst. Geol. Publ. Hungar.* **54**, 369–382.

Trusheim, F. 1957. Halokinese und strukturelle Entwicklung Norddeutschlands. *Zeitschr. deutsch. geol. Gesell.* **109**, 111–151.

Uhlig, V. 1911. Die marinen Reiche des Jura und der Unterkreide. *Mitt. geol. Gesell. Wien* **4**, 329–448.

Urlichs, M. 1966. Zur Fossilführung und Genese des Feuerlethens, der Rät–Lias–Grenzschichten und des unteren Lias bei Nürnberg. *Erlanger Geol. Abh.* **64**, 1–42.

Vakhrameev, V. A. 1964. Jurassic floras of the Indo-European and Siberian botanical–geographical regions. *Publ. Inst. Grand-Ducal Sect. Sci. Nat., Phys. Math. Luxembourg*, 411–421.

Vakhrameev, V. A. 1965. Jurassic floras of the U.S.S.R. *Palaeobotanist* **14**, 118 123.

Valencio, D. A. and Vilas, J. F. 1970. Palaeomagnetism of some Middle Jurassic lavas from south-east Argentina. *Nature* **225**, 262–264.

Valentine, J. W. 1971. Plate tectonics and shallow marine diversity and endemism, an actualistic model. *System. Zool.* **20**, 253–264.

Valentine, J. W. 1972. Conceptual models of ecosystem evolution. In: T. J. M. Schopt (editor), *Models in Paleobiology*, Freeman & Cooper, S. Francisco, pp. 192–216.

Valentine, J. W. and Moores, E. M. 1972. Global tectonics and the fossil record. *J. Geol.* **80**, 167–184.

Van Straaten, L. M. J. U. 1971. Origin of Solnhofen Limestone. *Geol. Mijnb.* **50**, 3–8.

Veevers, J. J. and Wells, A. T. 1961. The geology of the Canning Basin, Western Australia. *Australia, Bur. Mineral. Resources, Geol. Geophys. Bull.* **60**.

Vereshchagin, V. N. and Ronov, A. B. 1968 (editors). *Atlas of the Lithological–Paleogeographical maps of the U.S.S.R. Vol. 3. Triassic, Jurassic and Cretaceous.* Govt. Press, Moscow.

Verma, H. M. and Westermann, G. E. G. 1973. The Tithonian (Jurassic) ammonite fauna and stratigraphy of Sierra Catorce, San Luis Potosi, Mexico. *Bull. Am. Paleont.* **63**, 107–320.

Veselinović, D. 1963. La biostratigraphie du facies récifal du Jurassique supérieur dans la zone carpato–balkanique et le parallèle avec la zone dinarique intérieure (Serbie). *Assoc. Geol. Carpato-Balkan. 5th Cong. Vol.* **3**, Part 2, Bucharest, 295–306.

Viniegra, F. 1971. Age and evolution of salt basins of southeastern Mexico. *Bull. Am. Ass. Petrol. Geol.* **55**, 478–494.

Visser, W. A. and Hermes, J. J. 1962. Geological results of the exploration for oil in Netherlands New Guinea. *Verhandel. Koninkl. Ned. Geol. Mijnbouwk. Genook*, **20**.

Vossmerbäumer, H. 1970. Untersuchungen zur Bildungsgeschichte des Unteren Lias in Schonen (Schweden). *Geologica et Palaeontologica* **4**, 167–193.

Wall, D. 1965. Microplankton, pollen and spores from the Lower Jurassic in Britain. *Micropaleont.* **11**, 151–190.

References

Watt, W. S. 1969. The crust-parallel dyke swarm of south west Greenland in relation to the opening of the Labrador Sea. *Canad. J. Earth Sci.* **6**, 1320–1321.

Weber, H. S. 1964. Zur Stratigraphie und Ammonitenfauna des Braunjura (Dogger) der östlichen Schwäbischen Alb. *Arb. Geol. Paläont. Inst. TH Stuttgart* N.S. **44**, 1–174.

Wells, J. W. 1967. Corals as bathometers. *Mar. Geol.* **5**, 349–366.

Wendt, J. 1964. Stratigraphisch–paläontologische Untersuchungen im Dogger Westsiziliens. *Boll. Soc. paleont. Ital.* **2**, 57–145.

Wendt, J. 1969. Stratigraphie und Paläogeographie des roten Jurakalks im Sonnwendgebirge (Tirol Österreich). *N. Jb. Geol. Paläont. Abh.* **132**, 219–238.

Wendt, J. 1970. Stratigraphische kondensation in triadischen und jurassischen Cephalopodenkalken der Tethys. *N. Jb. Geol. Paläont. Mh.* **7**, 433–448.

Wendt, J. 1971a. Genese und Fauna submariner sedimentarer Spaltenfullungen im Mediterranen Jura. *Palaeontographica* **136**, 121–192.

Wendt, J. 1971b. Die Typolokalität der Adneterschichten (Lias Österreich). *Ann. Inst. Geol. Publ. Hungar.* **54**, 105–116.

Wesley, A. 1973. Jurassic plants. In: A. Hallam (editor), *Atlas of Palaeobiogeography*, Elsevier, Amsterdam, pp. 329–338.

West, I. M. 1964. Evaporite diagenesis in the Lower Purbeck beds of Dorset. *Proc. Yorks. geol. Soc.* **34**, 315–330.

West, I. M. 1965. Macrocell structure and enterolithic veins in British Purbeck gypsum and anhydrite. *Proc. Yorks. geol. Soc.* **35**, 47–58.

Westermann, G. E. G. 1964. The ammonite fauna of the Kialagvik formation at Wide Bay, Alaska Peninsula. Part 1 – Lower Bajocian. *Bull. Am. Paleont.* **47**, 327–503.

Westermann, G. E. G. 1967. Sucesion de ammonites del Jurassico medio en Antofagasta, Atacama, Mendoza y Neuquen. *Rev. Assoc. geol. Argentina* **22**, 65–73.

Westermann, G. E. G. 1969. The ammonite fauna of the Kialagvik formation at Wide Bay, Alaska Peninsula. Part 2. Sonninia sowerbyi zone (Bajocian). *Bull. Am. Paleont.* **57**, 5–226.

Westermann, G. E. G. and Getty, T. A. 1970. New Middle Jurassic Ammonitina from New Guinea. *Bull. Am. Paleont.* **57**, 231–321.

Westermann, G. E. G. and Riccardi, A. C. 1972. Amonitas y estratigrafia del Aalenio – Bayociano en los Andes Argentina – Chilenos. *Ameghiniana* **9**, 357–389.

Wetzel, R. and Morton, D. M. 1958. Contribution à la Géologie de la Transjordanie. *Notes Mém. Moyen-Orient.* **7**, 95–191.

Whittaker, A. 1972. Intra-Liassic structures in the Severn Basin area. *I.G.S. Rep.* 72/3.

Wiedenmayer, F. 1963. Obere Trias bis mittleren Lias zwischen Saltrio und Tremona (Lombardische Alpen). Die Wechselbeziehungen zwischen Stratigraphie, Sedimentologie und syngenetischer Tektonik. *Ecl. geol. Helv.* **56**, 529–640.

Wierzbowski, A. 1966. L'Oxfordien supérieur et le Kimmeridgian inférieur du Plateau de Wielun. *Acta Geol. Pol.* **16**, 127–200. (In Polish, with French summary.)

Williams, M. D. 1959. Stratigraphy of the Lower Indus basin, West Pakistan. *Proc. 5th World Petrol. Cong.* Sect. 1, 377–394.

257

References

Wilson, R. C. C. 1968a. Upper Oxfordian palaeogeography of southern England. *Palaeogeog., Palaeoclimatol., Palaeoecol.* **4**, 5–28.

Wilson, R. C. L. 1968b. Carbonate facies variation within the Osmington Oolite Series in Southern England. *Palaeogeog. Palaeoclimatol., Palaeoecol.* **4**, 89–123.

Wilson, V., Welch, F. B. A., Robbie, J. A. and Green, G. W. 1958. *Geology of the country around Bridport and Yeovil.* Mem. geol. Surv. G.B.

Wincierz, J. 1973. Küstensedimente und Ichnofauna aus dem oberen Hettangium von Mackendorf (Niedersachsen). *N. Jb. Geol. Paläont. Abh.* **144**, 104–141.

Wobber, F. J. 1965. Sedimentology of the Lias (Lower Jurassic) of South Wales. *J. sedim. Petrol.* **35**, 683–703.

Wolfenden, B. 1965. Bau mining district, west Sarawak, Malaysia. Part 1: Bau. *Geol. Surv. Borneo. Bull.* **7**.

Woodland, A. W. 1971 (editor). The Llanbedr (Mochras Farm) borehole. *I.G.S. Rep.* 71/18.

Woodwell, G. M. and Smith, H. H. (eds.). 1969. *Diversity and stability in Ecological Systems.* Brookhaven Sympos. Biol. 22.

Worssam, B. C. and Ivimey-Cook, H. C. 1971. The stratigraphy of the Geological Survey borehole at Warlingham, Surrey. *Bull. geol. Surv. G.B.* No. 36, 1–144.

Wright, J. K. 1968. The stratigraphy of the Callovian rocks between Newtondale and the Scarborough coast, Yorkshire. *Proc. Geol. Ass. Lond.* **79**, 363–399.

Wright, J. K. 1972. The stratigraphy of the Yorkshire Corallian. *Proc. Yorks geol. Soc.* **39**, 225–264.

Yeats, R. S. 1968. Southern California structure, sea floor spreading and history of the Pacific basin. *Bull. Geol. Soc. Am.* **79**, 1693–1702.

Yen, T. C. 1950. Molluscan fauna of the Morrison formation. U.S. Geol. Surv. Prof. Paper 233B, 21–51.

Zanzucchi, G. 1963. Le ammoniti del Lias superiore (Toarciano) di Entratico in Vel Cavallina (Bargamasco orientale). *Mem. Soc. Ital. Sc. nat. Milano* **13**, 101–146.

Zeiss, A. 1968. Untersuchungen zur Paläontologie der Cephalopoden der Unter-Tithon der südlichen Frankenalb. *Bay. Akad. Wiss. Math.–Naturwiss. Kl. Abh.* N.S. **132**.

Ziegler, B. 1961. Stratigraphische und zoogeographische Beobachtungen an *Aulacostephanus* (Ammonoidea – Oberjura). *Paläont. z.* **35**, 79–89.

Ziegler, B. 1963. Die Fauna der Lemeš-Schichten (Dalmatien) und ihre Bedeutung für den mediterranen Oberjura. *N. Jb. Geol. Paläont. Mh.* **8**, 405–421.

Ziegler, B. 1964. Boreal Einflüsse im Oberjura Westeuropas? *Geol. Rundsch.* **54**, 250–261.

Ziegler, B. 1967. Ammoniten-Ökologie am Beispiel des Oberjura. *Geol. Rundsch.* **56**, 439–464.

Ziegler, P. A. 1956. Zur stratigraphie des Sequanien im zentralen schweizer Jura. *Beitr. Geol. Schweiz* N.S. **102**, 37–101.

Index

Index

Index

Index

Index

Index